Modern Cabinetmaking

SIXTH EDITION

Brian Lee Skates

Nancy Henke-Konopasek

Publisher

The Goodheart-Willcox Company, Inc.

Tinley Park, IL

www.g-w.com

Introduction

This lab workbook is designed for use with the text, **Modern Cabinetmaking**. As you complete the questions and problems in this workbook, you can review the facts and concepts presented in the text.

Modern Cabinetmaking is a comprehensive text that explains and illustrates all aspects of fine woodworking. You will study furniture styles and designs that will support an orderly approach to the design/construction process. You will learn how to prepare sketches or drawings and how to prepare a plan of procedure before beginning work.

Modern Cabinetmaking also provides you with a thorough understanding of the technology of woodworking. You will learn the "How, What, Why, and When" of selecting appropriate materials whether they are fine hardwoods, manufactured panel products, plastics, glass, or ceramics. Hardware, abrasives, adhesives, and finishing supplies are discussed in detail.

Modern Cabinetmaking provides complete coverage of processes and techniques using both hand and power tools and machines. As the book guides you through the processes, it shows you how your decisions will affect the final product. As you study and practice your techniques, you will increase your woodworking and cabinetmaking skills. You will learn to reach make-buy decisions based on your own experience and skill. You will understand how the availability of hand, portable, and stationary tools and machines influences your decisions.

Whether you work at home, in industry, or in a school, your activities should focus on your own safety, the safety of your coworkers, and the effects of the process on the environment. Always try to protect yourself and others around you against the risk of accidents and injuries.

Lab workbook chapters correspond to chapters in the text. After reading your assignment in the text, do your best to complete the questions and problems carefully and accurately. The lab workbook chapters also contain activities related to textbook chapter content. The activities range from chapter content reinforcement to real-world application. It is important in these activities to understand any safety procedures set by your teacher.

Section Projects are to be completed after studying the textbook sections. The Section Projects allow you to apply the ideas, theories, and concepts learned in the textbook. The Section Projects will incorporate these ideas to help you improve your hands-on techniques and skills.

This lab workbook helps to provide the foundation on which a sound, thorough knowledge of cabinetmaking is based. This lab workbook has been designed to increase your interest and understanding of the text material.

The content of this Lab Workbook and the text correlate to Woodwork Career Alliance (WCA) skill standards. The WCA establishes a benchmark to measure and recognize an individual's skills and knowledge. The WCA skill standards help ensure that students are prepared for rigorous industry standards, and provide a pathway for advancement for professional woodworkers.

Contents

SECTION 1
Industry Overview

Chapter 1 Introduction to Cabinetmaking 1
Chapter Review 1
Activity 1A: Design Your Company Layout 3

Chapter 2 Health and Safety 5
Chapter Review 5
Activity 2A: Shop Safety Rules 8

Chapter 3 Career Opportunities 9
Chapter Review 9
Activity 3A: Career Investigation 12
Activity 3B: Preparing a Résumé 14
Activity 3C: Finding Employment 15
Activity 3D: Completing a Job Application 16
Activity 3E: The Job Interview 19
Activity 3F: Entrepreneurship 21

Chapter 4 Cabinetmaking Industry
Overview 23
Chapter Review 23
Activity 4A: Industry Apprenticeship 25
Activity 4B: You, the Cabinetmaking
Professional 26

SECTION 2
Design and Layout

Chapter 5 Cabinetry Styles 27
Chapter Review 27
Activity 5A: Stylish Table and Chair 33

Chapter 6 Components of Design 35
Chapter Review 35
Activity 6A: Principles of Design 38
Activity 6B: The Golden Mean 39

Chapter 7 Design Decisions 41
Chapter Review 41
Activity 7A: Cabinet Design 43

Chapter 8 Human Factors 45
Chapter Review 45
Activity 8A: Human Factors and Safety 47
Activity 8B: Designing for Children 48

Chapter 9 Production Decisions 49
Chapter Review 49
Activity 9A: Leaning Bookshelf 52

Chapter 10 Sketches, Mock-Ups, and Working
Drawings 55
Chapter Review 55
Activity 10A: Bookshelf 58

Chapter 11 Creating Working Drawings 59
Chapter Review 59
Activity 11A: Alphabet of Lines 62
Activity 11B: Lettering 63

Chapter 12 Measuring, Marking, and Laying Out
Materials 65
Chapter Review 65
Activity 12A: Transferring a Design 72

SECTION 3
Materials

Chapter 13 Wood Characteristics 73
Chapter Review 73
Activity 13A: Chart of Wood Characteristics 78

Chapter 14 Lumber and Millwork 79
Chapter Review 79
Activity 14A: Specialty Item 83

Chapter 15 Cabinet and Furniture Woods 85
Chapter Review 85
Activity 15A: Wood Characteristics 87
Activity 15B: What Wood to Use? 92

Chapter 16 Manufactured Panel Products 93
Chapter Review 93
Activity 16A: Choosing the Right Material for the
Job 96

Chapter 17 Veneers and Plastic Overlays 97
Chapter Review 97
Activity 17A: Material Warpage 102

Chapter 18 Glass, Plastic, and Related
Products 103
Chapter Review .103
Activity 18A: Countertop Materials 107

Chapter 19 Hardware 109
Chapter Review .109
Activity 19A: Which Way to Open? 112

Chapter 20 Fasteners 113
Chapter Review .113
Activity 20A: Purchasing Agent 118

Chapter 21 Ordering Materials and Supplies . . .119
Chapter Review .119
Activity 21A: Material and Supplies Order Form . . .121

SECTION **4**
Machining Processes

Chapter 22 Sawing with Hand and Portable
Power Tools 123
Chapter Review .123
Activity 22A: Which Tool to Use? 127

Chapter 23 Sawing with Stationary Power
Machines 129
Chapter Review .129
Activity 23A: Using Stationary Power Tools 136

Chapter 24 Surfacing with Hand and Portable
Power Tools 139
Chapter Review .139
Activity 24A: The Hand Plane 143

Chapter 25 Surfacing with Stationary
Machines 145
Chapter Review .145
Activity 25A: Troubleshooting and Correcting
Planer Problems 150

Chapter 26 Shaping 153
Chapter Review .153
Activity 26A: Shaping an Edge 158

Chapter 27 Drilling and Boring 161
Chapter Review .161
Activity 27A: Cabinet Installer 166

Chapter 28 Computer Numerically Controlled
(CNC) Machinery 167
Chapter Review .167
Activity 28A: Figuring CNC Coordinates 171

Chapter 29 Abrasives 173
Chapter Review .173

Chapter 30 Using Abrasives and Sanding
Machines 177
Chapter Review .177
Activity 30A: Setting Up Shop with Abrasive
Machines 180

Chapter 31 Adhesives 181
Chapter Review .181
Activity 31A: Supply the Adhesive 186

Chapter 32 Gluing and Clamping 187
Chapter Review .187

Chapter 33 Bending and Laminating 193
Chapter Review .193
Activity 33A: Designing for Bending and
Laminating 196

Chapter 34 Overlaying and Inlaying Veneer 197
Chapter Review .197
Activity 34A: Veneer Chess Board 202

Chapter 35 Installing Plastic Laminates 203
Chapter Review .203
Activity 35A: Laminating an Island Top 205

Chapter 36 Turning 207
Chapter Review .207
Activity 36A: Turning Projects 214

Chapter 37 Joinery 215
Chapter Review .215
Activity 37A: Choose a Joint 223

Chapter 38 Accessories, Jigs, and Special
Machines 225
Chapter Review 225
Activity 38A: Shop Accessories, Jigs, and Special
Machines 229

Chapter 39 Sharpening 231
Chapter Review 231
Activity 39A: Shaping an Edge 233

SECTION 5
Fabrication and Installation

Chapter 40 Case Construction 235
Chapter Review 235
Activity 40A: Casework Joints 239

Chapter 41 Frame and Panel Construction 241
Chapter Review 241
Activity 41A: Frame-and-Panel Door Design 245

Chapter 42 Cabinet and Furniture Supports . . .247
Chapter Review 247
Activity 42A: Choose Your Support 251

Chapter 43 Doors. 253
Chapter Review 253
Activity 43A: Door Design Decisions. 258

Chapter 44 Drawers 259
Chapter Review 259
Activity 44A: Your Drawer Design 264

Chapter 45 Cabinet Tops and Tabletops 267
Chapter Review 267
Activity 45A: Tabletop. 272

Chapter 46 Kitchen Cabinet Design and
Installation 273
Chapter Review 273
Activity 46A: Kitchen Cabinet Checklist. 279
Activity 46B: Measuring Appliances. 280

Chapter 47 Built-in Cabinetry, Wall Paneling,
and Mantels. 283
Chapter Review 283
Activity 47A: Add Built-in Cabinets to Your
Home 287

Chapter 48 Installing Moulding and Trim 289
Chapter Review 289
Activity 48A: Running Base 293

Chapter 49 Furniture Project Ideas. 295
Chapter Review 295
Activity 49A: Selecting Furniture 298

SECTION 6
Finishing

Chapter 50 Finishing Decisions 299
Chapter Review 299
Activity 50A: Decisions! Decisions! Decisions! . . . 302

Chapter 51 Preparing Surfaces for Finish 303
Chapter Review 303
Activity 51A: Which Tool to Use?. 306

Chapter 52 Finishing Tools and Equipment . . . 307
Chapter Review 307
Activity 52A: Finish a Piece in Your Garage. . . . 312
Activity 52B: Set Up a Finishing Department . . . 313

Chapter 53 Stains, Fillers, Sealers, and Decorative
Finishes 315
Chapter Review 315
Activity 53A: Choose a Decorative Finish 318

Chapter 54 Topcoatings 319
Chapter Review 319
Activity 54A: Choose a Topcoating 322

Section Project 1-1 Class PowerPoint®
Presentation 323

Section Project 1-2 Mini Flip Chart of
Professional Organizations 326

Section Project 1-3 Identifying Shop Hazards. . . 329

Section Project 2-1 Project Management 333

Section Project 2-2 Student Desk and Desk
Chair Design 344

Section Project 2-3 Components of Design
Poster 349

Section Project 3-1 Wood Samples Stringer. . . 353

Section Project 3-2 Moisture Content Material
Testing. 356

Section Project 4-1 Making a Bench Hook. . . . 361

Section Project 4-2 Drill Press Practice. 368

Section Project 4-3 Dry Fit. 373

Section Project 4-4 Building a Machine Jig or Fixture. 376

Section Project 4-5 Wood Joints Class Project. . . 384

Section Project 4-6 Make a Push Stick from a Pattern 390

Section Project 4-7 Make a Routed Sign 393

Section Project 4-8 Make a Sanding Block. . . . 400

Section Project 4-9 Build a Sawhorse 403

Section Project 4-10 Sharpening a Chisel. 408

Section Project 4-11 Make a Sign with a CNC Router 411

Section Project 4-12 Veneering 414

Section Project 4-13 Edge Trim 417

Section Project 4-14 Practice Applying Plastic Laminate 421

Section Project 5-1 Surfacing Stock for Cabinet Face Frames. 425

Section Project 5-2 Trinket Box 428

Section Project 5-3 Bedside Face Frame Cabinets. 435

Section Project 5-4 Wall Cabinets—Face Frame and Carcase Construction 443

Section Project 5-5 Frameless Bedside Stand Cabinets. 452

Section Project 5-6 Frameless Wall Cabinets . . . 460

Section Project 6-1 Fixing a Dent 469

Section Project 6-2 Finish Sample Set 472

CHAPTER 1

Introduction to Cabinetmaking

Chapter Review

Carefully read Chapter 1 of the text and answer the following questions.

1. List three examples of wood products used in everyday life.

   _____

   _____

   _____

2. Conclusions made about a product design before work begins are known as _____.

   _____

3. List two factors on which all design decisions are based.

   _____

4. The purpose for having a cabinet or piece of furniture is known as _____.

   _____

5. The appearance of a cabinet is known as _____.

   _____

6. Production cabinet shops often use _____ systems to create their designs.

   _____

7. _____ An example of a type of cabinetry designed and produced based on standards would be _____.
   A. factory-produced kitchen cabinets
   B. custom cabinets
   C. ready-to-assemble cabinetry
   D. None of the above.

8. _____ Which of the following describes ready-to-assemble cabinetry?
   A. The finished cabinet is purchased unassembled in a neatly packaged, compact carton.
   B. The consumer assembles the cabinet with special RTA fasteners.
   C. Large furniture can be moved easily because it can be disassembled and reassembled in a new location.
   D. All of the above.

9. Describe material decisions that you would need to make if you wanted to produce a cabinet.

   _____

   _____

   _____

   _____

   _____

   _____

10. _____ _____ is a sheet of thinly sliced wood used to cover poor-quality lumber.
    A. Particleboard
    B. Plywood
    C. Veneer
    D. Fiberglass

11. _____ decisions include choosing the tools and procedures necessary to build a product in the most efficient manner.

_____

12. What is tooling?

_____

_____

_____

13. _____ refers to cutting or removing material.

_____

14. _____ includes all operations where material is bent into a shape using a mold.

_____

15. _____ involves assembling or joining two materials.

_____

16. _____ Making a mock-up is an example of which type of activity?
    A. Processing
    B. Preprocessing
    C. Post-processing
    D. None of the above.

17. _____ includes all tasks from cutting standard stock to finishing the product.

_____

18. Transporting, installing, and maintaining are all _____ activities.

_____

19. List four activities management needs to do in order to accomplish a task.

_____

_____

_____

_____

20. _____ Which of the following statements regarding quality is *true*?
    A. It can be specified by the designer or the user.
    B. It is measured by how well a product meets a consumer's requirements and expectations.
    C. Standards for quality have been established and documented.
    D. All of the above.

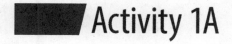 **Activity 1A**

Design Your Company Layout

Imagine it is 10 to 15 years from now and you have decided to start your own cabinetry company. Use the space provided to draw a simple floor plan, labeling where all the elements of your company would function. Most cabinetry companies have an area for the office, production, receiving supplies, shipping products, etc. Review the chapter to place the other elements. Focus more on including all the elements required and on the flow of cabinetry throughout your company rather than on the actual size of your company.

Notes

_____

_____

_____

_____

_____

_____

_____

_____

_____

_____

_____

_____

_____

_____

_____

_____

_____

_____

_____

_____

_____

_____

_____

_____

_____

_____

_____

_____

_____

_____

_____

CHAPTER 2 — Health and Safety

Chapter Review

Carefully read Chapter 2 of the text and answer the following questions.

1. _____ Accidents occur as the result of _____.
   A. hazardous conditions
   B. reading labels, safety instructions, and caution and warning signs
   C. unsafe acts
   D. Both A and C.

2. Why was the Occupational Safety and Health Administration (OSHA) established?

   _____

   _____

3. _____ The best safety preparation for any cabinetmaker is to _____.
   A. read
   B. watch
   C. understand
   D. All of the above.

4. Mark each statement that describes an unsafe act.

   A. _____ Performing a machine operation with proper knowledge and thorough planning.

   B. _____ Working while under physical or emotional stress.

   C. _____ Taking breaks when frustration interferes with concentration.

   D. _____ Distracting someone who is operating a machine.

   E. _____ Making setups when the machine is running.

   F. _____ Putting all guards in their protective positions.

   G. _____ Carrying sharp tools only by the handle.

   H. _____ Lifting with the back.

5. List three ways to reduce the slippery nature of concrete or wood floors.

   _____

   _____

   _____

6. Class 1 liquids are the most dangerous because the _____ is within room temperature.

   _____

7. _____ Store flammable liquids in _____.
   A. old plastic soft drink bottles
   B. glass jars
   C. approved steel safety cans
   D. ungrounded metal containers

8. List three examples of safe material storage.

_____

_____

_____

_____

_____

9. Machines and electrical equipment should be wired in compliance with the _____.

_____

10. Electrical _____ can prevent you from being shocked or electrocuted.

_____

11. Air hoses used for _____ removal have a pressure relief nozzle.

_____

12. _____ *True or False?* OSHA requires written lockout/tagout programs for machines with two or more energy sources.

13. Name three types of fire protection devices.

_____

_____

_____

14. _____ Fire extinguishers should be located within _____' of any work area.
   A. 25
   B. 50
   C. 75
   D. 100

15. _____ A pressurized water type extinguisher is used to extinguish Class _____ fires.
   A. A
   B. B
   C. C
   D. D

16. _____ The carbon dioxide extinguisher is effective on Class _____ fires.
   A. A and B
   B. B and C
   C. C and D
   D. D

17. List and describe five types of personal protective equipment that should be worn in the shop.

_____

_____

_____

_____

_____

_____

_____

18. _____ *True or False?* Gloves should be worn while operating power machinery.

19. _____ _____ guards prevent machines from operating while dangerous parts are exposed.
    A. Automatic
    B. Interlocking
    C. Enclosure
    D. Point-of-operation

20. Why should people who work with woodworking machinery have first-aid training?

    _____

    _____

    _____

Match each item to the statement that describes it.

21. _____ Device that will automatically disconnect an electric circuit if a problem exists in an electric tool or wiring.

22. _____ Feature that keeps a tool running even when the hand is removed.

23. _____ Removes dust particles from machines.

24. _____ Details the properties and hazards of chemical products.

25. _____ Provides two layers of insulation.

26. _____ Wire screen placed in the neck of a can that prevents flames from getting inside the can.

A. Dust collection system
B. Safety data sheet
C. Flame arrestor
D. Trigger lock
E. Ground fault circuit interrupter (GFCI)
F. Double-insulated tool

 Activity 2A

Shop Safety Rules

Make a list of safety rules that should be followed in your classroom shop. Your answers should indicate a thorough study of the chapter.

_____

_____

_____

_____

_____

_____

_____

_____

_____

_____

_____

_____

_____

_____

_____

_____

_____

_____

_____

_____

_____

_____

_____

_____

_____

_____

_____

_____

_____

_____

_____

CHAPTER
3
Career Opportunities

Chapter Review

Carefully read Chapter 3 of the text and answer the following questions.

1. _____ The cabinetmaking industry is part of the _____ wood products industry.
   A. primary
   B. secondary
   C. aftermarket
   D. decorative

2. _____ *True or False?* Generally speaking, cabinets are now built individually and directly on-site.

3. _____ Career advancement requires you to continue to _____.
   A. increase your knowledge of the industry
   B. improve your on-the-job skills
   C. maintain a positive attitude
   D. All of the above.

4. Provide two questions that ambitious individuals in the cabinetmaking industry might ask themselves.

   _____

   _____

   _____

   _____

5. What is the *Occupational Outlook Handbook*?

   _____

   _____

   _____

   _____

6. List three positions generally found within the management area in a business.

   _____

   _____

   _____

7. Positions in the design area require some _____ skills.

   _____

8. Name four ways to obtain training in the cabinetmaking industry.

   _____

   _____

   _____

   _____

9. Products that combine wood and other materials are known as _____.

_____

10. Altering the molecular structure of coatings to reduce the impact of ultraviolet radiation on wood is an example of _____.

_____

11. _____ Which of the following statements about apprenticeship is *true*?
   A. It offers the opportunity to get paid while you learn.
   B. It is not necessary to have an employer sponsor.
   C. After successful completion, the apprentice becomes a master craftsperson.
   D. It is not necessary to attend classes as part of the apprenticeship.

12. Communication, math, problem-solving, and leadership skills are known as _____ skills.

_____

13. List four resources to use to find employment.

_____

_____

_____

_____

14. Most employment begins with _____, usually done on the job.

_____

15. _____ *True or False?* Employers expect 100% productivity when you start a job.

16. _____ Quality work means _____.
   A. excellent product appearance
   B. proper tool maintenance
   C. holding to a production schedule
   D. All of the above.

17. Increased experience at the job and having a good attitude can result in _____ opportunities.

_____

18. _____ If you are resigning from your job, give advance notice of _____.
   A. one day
   B. two weeks
   C. one year
   D. six months

19. List four reasons an employee might be fired.

_____

_____

_____

_____

20. A(n) _____ is the potential for advancement in responsibility, job title, and pay.

_____

21. When building a career, join a professional _____ to observe the characteristics of people who have the type of jobs you would like.

_____

22. Identify three strategies for career advancement.

_____

_____

_____

_____

_____

_____

23. A(n) _____ is the highest-ranking corporate officer or administrator in charge of total management of an organization.

_____

24. A(n) _____ is someone who starts a business.

_____

25. A(n) _____ is made up of those individuals who will buy your product.

_____

26. What is consignment production?

_____

_____

_____

_____

27. A business specializing in furniture restoration and repairs is known as a(n) _____ service business.

_____

Name _____ Date _____ Class _____

 Activity 3A

Career Investigation

Investigate a career of your choice in the field of cabinetmaking. If possible, interview someone with that career. Conduct thorough research. Sources of information include the library, the Internet, and the guidance office at your school. Find the following information.

Career: _____

Education or training needed: _____

Salary range: _____

Description of duties: _____

_____

_____

_____

_____

Hours: _____

Working conditions: _____

_____

_____

_____

_____

Health and safety concerns: _____

_____

_____

_____

Advantages of this career: _____

_____

_____

_____

_____

Name _____

Disadvantages of this career: _____

_____

_____

_____

_____

Other comments: _____

_____

_____

_____

_____

 **Activity 3B**

Preparing a Résumé

Design your résumé in the space provided. When you have finished, carefully reread it. Ask your instructor or counselor to read it. When you are satisfied with your résumé, type it using desktop publishing software. You may wish to make copies of your résumé to submit to potential employers.

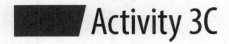 Activity 3C

Finding Employment

Find an advertisement for a cabinetmaking job on either a job search website or in your local newspaper. Place a copy of the advertisement in the space provided. Next, write an application letter for the job, using the space provided. Be sure to state in the letter what you can do for the company. Your application letter will help a prospective employer to form an impression of you.

Job Advertisement

Application Letter

Name _____ Date _____ Class _____

 Activity 3D

Completing a Job Application

Pretend that you are applying for the job described in the advertisement on the previous page. Complete the following job application.

Personal Information

Name: _____

Address: _____

Telephone: _____ E-mail address: _____

Job Interest

Position for which you are applying: _____

Salary desired: _____

Date you can start: _____

Have you ever been employed by this company before? _____

If yes, where? _____ When? _____

Education and Training

_____

_____

_____

_____

_____

_____

_____

_____

_____

_____

_____

_____

Name _____

Employment Record

_____

_____

_____

_____

References

List the names of three people not related to you whom you have known at least one year.

_____

_____

_____

Additional Data

Have you ever been convicted of a crime (other than traffic, game law, or other minor violations)? _____

If yes, explain the offense and circumstances regarding the conviction:

_____

Age: _____

Are you a US citizen? _____

If no, do you have an alien registration card or valid US work permit? _____

Non-English languages you read: _____

Speak: _____

Write: _____

Special skills, knowledge, and abilities that qualify you for the position you are seeking:

_____

_____

Health Information

Are you presently or have you during the last six months been under a physician's care or in a hospital? _____

Have you ever been compensated for, or do you currently have outstanding, a job-related claim?

_____

If you answered yes to any of the above, please explain.

_____

_____

Thank you for completing this application and for your interest in employment with us. We would like to assure you that your opportunity for employment with this company will be based on your merit only, without regard to your race, religion, sex, age, national origin, or disability.

Please read carefully:

Applicant's Certification and Agreement

I certify that the facts contained in this application are true and complete to the best of my knowledge. I understand that, if employed, falsified statements on this application shall be grounds for dismissal. I authorize investigation of all statements contained in this application. I authorize the references listed above to give you any and all information concerning my previous employment and any pertinent information they may have, personal or otherwise. I release all parties from all liability for any damage that may result from furnishing this information to you.

Signature _____   Date _____

Name _____ Date _____ Class _____

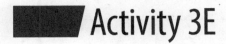 Activity 3E

The Job Interview

Commonly asked interview questions are listed. Pretend that you are being interviewed. Answer these questions as you would during an interview.

Job for which you are applying: _____

Company: _____

Why do you want to work for this company? _____

_____

_____

_____

Do you think you will like this kind of work? _____ Why? _____

_____

_____

How would you describe yourself as a worker? _____

_____

_____

_____

_____

What are your best subjects in school? _____

_____

What are your worst subjects in school? _____

_____

What other jobs have you had? _____

_____

_____

_____

Have you ever been fired from a job? _____ If so, why? _____

_____

_____

_____

What is your best qualification for this job? _____

_____

_____

_____

What are your future plans? _____

_____

Why should we hire you? _____

_____

_____

_____

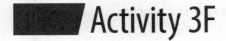 Activity 3F

Entrepreneurship

Complete the following activity about entrepreneurship.

1. Define entrepreneurship in your own words.

   _____

   _____

   _____

2. Does starting your own business appeal to you? Give two reasons for your answer.

   _____

   _____

   _____

3. List at least three products or services your business could sell.

   _____

   _____

   _____

4. Briefly describe the demand for your products or services. Consider potential customers, competition, and your abilities.

   _____

   _____

   _____

5. Based on your answer to the previous question, which product or service seems to offer you the best opportunity for success?

   _____

6. How much time would your business take to manage? Do you have enough time to accomplish this?

   _____

   _____

   _____

7. What do you estimate it would cost to start and operate your business? Consider equipment, supplies, advertising, and other operating costs.

   _____

   _____

8. Consider your answers to the previous question. Based on your answers, list several advantages and disadvantages of becoming an entrepreneur.

_____

_____

_____

_____

CHAPTER 4
Cabinetmaking Industry Overview

Chapter Review

Carefully read Chapter 4 of the text and answer the following questions.

1. Cabinetmaking is one of several _____ in the secondary wood products industry.

   _____

2. List four subsectors in the secondary wood products industry.

   _____

   _____

   _____

   _____

3. _____ is custom designed and manufactured with a high degree of quality and precision.

   _____

4. Kitchen and bath cabinets are examples of _____ cabinetry.

   _____

5. What is meant by *reshoring*?

   _____

   _____

6. _____ *True or False?* Building one-of-a-kind products is known as contract cabinetmaking.

7. _____ Which of the following statements regarding batch production is *not true*?
   A. A fixed quantity of units is produced at one time.
   B. It must be well-timed.
   C. One company is hired by another company to make products.
   D. It is a standard practice for building stock products.

8. _____ The _____ zone includes the offices of company executives, buyers, and production engineers.
   A. active
   B. quiet
   C. support
   D. shipping

9. What is *lean manufacturing*?

   _____

   _____

   _____

   _____

10. What are two advantages of using *just-in-time (JIT)* production?

_____

_____

_____

11. List seven benefits that trade associations provide.

_____

_____

_____

_____

_____

_____

_____

12. The _____ has developed an extensive set of standards and a credentialing system that provides a path for individuals to document their skills.

_____

13. _____ Which trade organization offers a professional certification program that teaches individuals how to operate an ethical, sustainable, and profitable custom woodworking enterprise from an owner's or manager's perspective?
   A. Forest Products Society
   B. Woodwork Career Alliance
   C. Cabinet Makers Association
   D. Woodwork Institute

14. _____ provide a great opportunity to see new machinery and products.

_____

Match each trade show with the correct description.

15. _____ Trade event dedicated to kitchen and bath industry.

16. _____ Woodworking professionals gather to learn how to keep their businesses productive, competitive, and cutting edge.

17. _____ Largest trade show in the world dedicated to wood products and processing.

18. _____ Largest trade show for the wood industry in North America.

A. International Wood Fair (IWF)

B. Ligna

C. Association of Woodworking and Furnishings Suppliers (AWFS) Fair

D. Kitchen and Bath Industry Show (KBIS)

Name _____ Date _____ Class _____

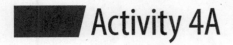 **Activity 4A**

Industry Apprenticeship

Assume you have gotten your first job as a cabinetmaking apprentice, and then answer the following questions.

1. What trade association would you join first? Why?

_____

_____

_____

_____

_____

_____

_____

2. What shows would you attend? Why?

_____

_____

_____

_____

_____

_____

_____

3. Would you order a trade journal subscription? Why or why not?

_____

_____

_____

_____

_____

_____

_____

Activity 4B

You, the Cabinetmaking Professional

Imagine yourself in 10 to 20 years. You are a cabinetmaking professional. You own a cabinet shop or are the Chief Executive Officer (CEO) of a large cabinetmaking firm. In the space provided, write a report describing yourself and your responsibilities. What professional organizations do you belong to and why do you think they are important? What publications do you subscribe to and why are they important? What trade shows and conferences do you attend?

_____

_____

_____

_____

_____

_____

_____

_____

_____

_____

_____

_____

_____

_____

_____

_____

_____

_____

_____

_____

_____

_____

_____

_____

_____

_____

_____

_____

_____

CHAPTER 5 Cabinetry Styles

Chapter Review

Carefully read Chapter 5 of the text and answer the following questions.

1. _____ refers to the features of a cabinet that distinguish it from other pieces.

   _____

2. _____ *True or False?* Provincial furniture has more carving than traditional pieces.

3. _____ _____ furniture uses straight lines with very little carving.
   A. American Modern
   B. French Provincial
   C. William and Mary
   D. Contemporary

4. Place the following cabinetry styles in order, according to when they were developed: provincial, contemporary, traditional.

   _____

5. Explain how early cabinetmaking differs from modern cabinetmaking.

   _____

   _____

   _____

   _____

   _____

6. Name five traditional styles of cabinetry.

   _____

   _____

   _____

   _____

7. Identify the following style of furniture.

_____

L.W. Crossan Cabinetmaker; David Gentry Photographer

8. _____ A _____ leg is a curved leg that ends with an ornamental foot.
   A. contemporary
   B. cabriole
   C. fretwork
   D. lattice

9. Identify the following style of furniture.

_____

L.W. Crossan Cabinetmaker; David Gentry Photographer

10. _____ Which of the following statements regarding Chippendale-style furniture is correct?
    A. It often used veneer.
    B. It used inlays.
    C. It often included lattice work on table aprons.
    D. It had little Chinese influence.

Name _____

11. Identify the following style of furniture.

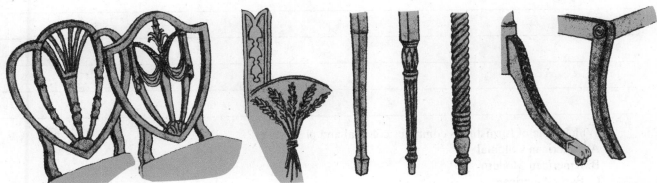

Colonial Homes

_____

12. _____ The secretary, a bookcase with a hinged front door, was introduced by _____.
   A. Thomas Sheraton
   B. George Hepplewhite
   C. Thomas Chippendale
   D. None of the above.

13. Simplified versions of European traditional styles were labeled _____.

_____

14. In early America, provincial style furniture was called _____.

_____

15. Name six provincial furniture styles.

_____

_____

_____

_____

_____

_____

Match each term with the correct description.

16. _____ Popular from mid 1600s to about 1790 and featured graceful, curved edges.

17. _____ Styling was crude, but some pieces were refined with European influences.

18. _____ Cabinets were often painted with flowers or animals.

19. _____ Extremely plain with very few decorations.

20. _____ Chairs were finely crafted by wheelwrights.

21. _____ Loosely followed Sheraton and Hepplewhite, but with more fine carving.

A. Shaker
B. American Colonial
C. Windsor
D. Pennsylvania Dutch
E. Duncan Phyfe
F. French Provincial

22. List five contemporary styles of cabinetry.

_____

_____

_____

_____

_____

23. _____ Which style of furnishings combines colonial and plain styles?
    A. American Colonial
    B. American Modern
    C. Early American
    D. Shaker Modern

24. _____ Clean, undecorated furniture is a characteristic of the _____ style.
    A. American Colonial
    B. American Modern
    C. Early American
    D. Shaker Modern

25. _____ Which of the following types of furniture often features an opaque lacquer surface?
    A. Shaker Modern
    B. Oriental Modern
    C. Scandinavian Modern
    D. None of the above.

26. _____ refers to how well a room matches the original historic design.

_____

27. Coordination of furniture styles creates _____ in the interior of the home.

_____

Identify the following styles of kitchens shown.

28. _____

Wood-Mode

29. _____

Wood-Mode

30. _____

Wood-Mode

31. _____

Wood-Mode

32. _____

Wood-Mode

33. _____

Wood-Mode

 **Activity 5A**

Stylish Table and Chair

What is your favorite furniture style? Sketch a simple table and chair in that style. Note: Your instructor is not looking for perfection of sketching technique here; you will develop that later. Your focus is to make the drawing communicate the specific elements that make your style distinct.

1. Describe the style you chose and why you like it.

_____

_____

_____

_____

_____

_____

Notes

Components of Design

Chapter Review

Carefully read Chapter 6 of the text and answer the following questions.

1. Name the two factors that guide the development of a product during design.

   _____

   _____

2. Give three examples of functional design in cabinetry.

   _____

   _____

   _____

3. _____ refers to the style and appearance of a product.

   _____

4. Name three different ability levels among designers.

   _____

   _____

   _____

5. _____ Lines, shapes, textures, and colors are the _____ of design.
   - A. forms
   - B. repetition
   - C. elements
   - D. intensity

6. _____ *True or False?* Straight lines and square corners are the easiest element to produce in cabinetry design.

7. Identify the following shapes shown.

   A. _____

   B. _____

   C. _____

   D. _____

   E. _____

   F. _____

   G. _____

   H. _____

   I. _____

   Goodheart-Willcox Publisher

8. Identify each of the following images as either a primary horizontal mass or a primary vertical mass.

A _____ B _____ *Thomasville Furniture Industries*

A. _____ B. _____

9. _____ is the pleasing relationship of all elements in a given product design.

_____

10. Using elements more than once to create a rhythm in a design is known as _____.

_____

11. _____ Which of the following is the correct definition of *formal balance*?
    A. It is a relationship between height and width.
    B. It is a relationship between height and length of a product.
    C. It is equal proportions on each side of a centerline in a design.
    D. None of the above.

12. A feeling of balance in a design, even though the design is not symmetrical, is known as _____.

_____

13. What is the value of the *golden mean*?

_____

14. _____ When designing a product, the designer must consider _____.
    A. the person who will use the product
    B. the elements and principles of design
    C. convenience and flexibility
    D. All of the above.

15. Give two examples each of convenience and flexibility in cabinet design.

_____

_____

_____

_____

Match each term with the correct definition.

16. _____ Brilliance of a color.

17. _____ Contain colored pigments that make an opaque finish.

18. _____ Red, yellow, and blue.

19. _____ The contour and feel of a product's surface.

20. _____ The lightness or darkness of a hue.

21. _____ Orange, green, and violet.

22. _____ Any color in its pure form.

23. _____ Color wood and enhance the grain pattern.

24. _____ Mixture of reds, yellows, and blues with oranges, greens, and violets.

A. Paints
B. Stains
C. Primary colors
D. Value
E. Intensity
F. Hue
G. Secondary colors
H. Tertiary colors
I. Texture

Name _____ Date _____ Class _____

 Activity 6A

Principles of Design

In the space provided, insert a picture from a magazine or newspaper showing a furniture setting featuring cabinetry. Then explain how each of the principles of design is achieved.

Repetition: _____

_____

_____

_____

Balance (formal balance): _____

_____

_____

_____

Balance (informal balance): _____

_____

_____

_____

Proportion: _____

_____

_____

_____

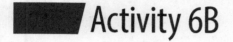 **Activity 6B**

The Golden Mean

In the space provided, draw three rectangles using the golden mean proportions. Draw one rectangle with the longest side at 3″ long, one with the longest side at 2″ long, and the third with the longest side at 1″ long. Round measurements to the nearest 1/16″ when calculating.

Notes

CHAPTER 7 / Design Decisions

Chapter Review

Carefully read Chapter 7 of the text and answer the following questions.

1. _____ The decision-making process _____.
   A. is inflexible
   B. commits you to one design idea
   C. guides your thoughts and actions during the many steps involved in creating a cabinet design
   D. All of the above.

2. Give an example of a *need* for a cabinet and a *want* for a cabinet.

   _____

   _____

   _____

   _____

3. _____ show how space will be arranged.

   _____

4. Look at the following two refined sketches. Identify the types of sketches they represent.

A

B   *O'Sullivan*

   A. _____   B. _____

5. _____ What type of design includes details?
   A. Preliminary
   B. Final
   C. Refined
   D. Working

6. _____ *True or False?* Dimensions are based on the intended function of a cabinet.

7. List five types of material classifications for cabinets.

_____

_____

_____

_____

_____

8. A(n) _____ contains all the materials that will be used to construct a cabinet.

_____

Match each term with the correct description.

9. _____ Contain all the drawings and specifications required to build a cabinet.

10. _____ Systematic process to determine the overall cost of equipment, materials, time, and space to build a product.

11. _____ Determines whether the product designed meets the needs of the user.

12. _____ Provide information on materials and tools needed to build a cabinet.

13. _____ Measures the ability of a material to withstand a load of force.

14. _____ Comparing advantages and disadvantages of making versus buying a component.

15. _____ Full-size working model of a design.

A. Strength test
B. Specifications
C. Cost analysis
D. Mock-up
E. Functional analysis
F. Make-or-buy decisions
G. Working drawings

Name _____ Date _____ Class _____

 **Activity 7A**

Cabinet Design

Identify and make a sketch of a cabinet or piece of furniture that you would like to produce. List five questions that you would ask and answer before you produce it.

Cabinet or piece of furniture: _____

_____

Sketch of the cabinet or piece of furniture:

Questions to ask and answer: _____

_____

_____

_____

_____

_____

Notes

CHAPTER 8

Human Factors

Chapter Review

Carefully read Chapter 8 of the text and answer the following questions.

1. List six human factors to consider when designing cabinetry.

   _____

   _____

   _____

   _____

2. _____ dimensions will meet the needs of a majority of people.

   _____

3. _____ It is wise to comply with standard dimensions unless _____.
   A. you are adapting for the elderly
   B. you want your design to be unique
   C. you are adapting for the disabled
   D. Both A and C.

4. Indicate the standard dimensions for kitchen cabinets using the image provided.

   A. _____

   B. _____

   C. _____

   D. _____

   E. _____

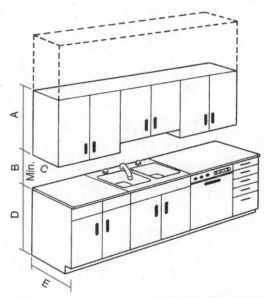

Goodheart-Willcox Publisher

5. _____ *True or False?* Counter heights may be standard or adapted to human factors.

6. _____ *True or False?* Machines, worktables, and walking space do not have standard dimensions associated with them.

7. What is the minimum required turning radius for wheelchairs?

_____

8. Seat heights are measured from the floor to the top of the _____ seat cushion.

_____

9. _____ Which type of chair is used for dining or working?
   A. Lounge
   B. Overstuffed
   C. Straight
   D. Side

10. _____ Standard seat height is _____".
   A. 17
   B. 18
   C. 19
   D. 20

11. _____ The difference between table height and compressed chair seat should be _____".
   A. 5
   B. 8
   C. 11
   D. 15

12. The area that a person can see is known as _____.

_____

13. _____ *True or False?* Sitting changes one's line of sight.

14. What type of safety hazard is presented by excessive stacking of items stored overhead?

_____

 Activity 8A

Human Factors and Safety

List two ways each of the following dangers can be eliminated or minimized in the design of furniture and cabinets.

1. Sharp edges and corners:

_____

_____

_____

_____

_____

_____

2. Falling objects:

_____

_____

_____

_____

_____

_____

3. Items within a child's reach:

_____

_____

_____

_____

_____

_____

_____

 **Activity 8B**

Designing for Children

In the space provided, draw cabinets similar to those shown in Figure 8-2 in the text. Change the dimensions to accommodate a 5-year-old child.

CHAPTER 9

Production Decisions

Chapter Review

Carefully read Chapter 9 of the text and answer the following questions.

1. Explain the purpose of production decisions.

   _____

   _____

   _____

2. List all the essential decision-making phases in the cabinetmaking process.

   _____

   _____

   _____

   _____

3. Writing a _____ prevents you from missing a crucial step.

   _____

4. In order, list the 11 general steps that should be included in a plan of procedure.

   _____

   _____

   _____

   _____

   _____

   _____

   _____

   _____

   _____

   _____

   _____

Match each term with the correct definition.

5. _____ Accessories used to perform cutting operations.

6. _____ Workpieces that are being processed into bill of material items.

7. _____ A device that holds a workpiece while the operator processes it.

8. _____ Material in its unprocessed form.

9. _____ Space made by the blade when cutting with a saw.

10. _____ Holds the workpiece and guides the tool.

11. _____ Rough cut stock that is ready to be sized.

A. Stock
B. Workpiece
C. Component
D. Tooling
E. Kerf
F. Fixture
G. Jig

12. _____ *True or False?* The type of tools you have affects the production time and the quality of the completed cabinet.

13. Give an example of a make-or-buy decision.

_____

_____

_____

14. What two items should you be aware of when laying out workpieces?

_____

_____

15. When cutting out materials, why should cuts be made from 1/32″ to 1/16″ (1 mm to 2 mm) oversize?

_____

_____

_____

16. _____ *True or False?* It is safe to assume that scales printed on machines are accurate.

17. On the following mortise and tenon joint shown, which should be cut first, the mortise or the tenon?

Tenon

Mortise

Goodheart-Willcox Publisher

_____

18. Holes, such as for installing hardware, are drilled _____ the stock is squared.

_____

19. What is often the most time-consuming part of the cabinetmaking process?

_____

Name _____

20. _____ Smoothing surfaces is done _____.
   A. before cutting and shaping each workpiece
   B. after cutting, but before shaping each workpiece
   C. after cutting and shaping each workpiece
   D. None of the above.

21. _____ The maximum time between spreading an adhesive and joining the components is known as _____ time.
   A. cure
   B. open
   C. clamp
   D. set

22. Assembling the product without adhesive is called a(n) _____.

_____

23. List four decisions that must be made before applying a finish to a product.

_____

_____

_____

_____

24. _____ *True or False?* Built-up finishes form a film on the surface of the wood.

25. List five methods for applying liquid coating materials.

_____

_____

_____

_____

_____

26. The last step in the cabinetmaking procedure is installing _____.

_____

◤ Activity 9A

Leaning Bookshelf

Make a plan of procedure for the production of the following leaning bookshelf project.

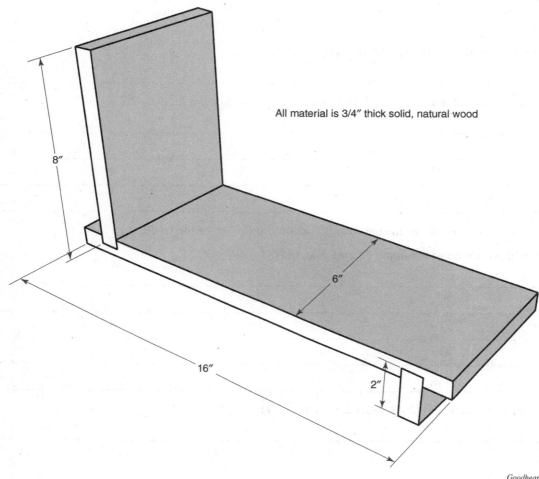

All material is 3/4″ thick solid, natural wood

8″

6″

16″

2″

Goodheart-Willcox Publisher

_____

_____

_____

_____

_____

_____

_____

_____

_____

_____

Name _____

_____

_____

_____

_____

_____

_____

_____

_____

_____

_____

_____

_____

_____

_____

_____

_____

_____

_____

_____

_____

_____

_____

_____

_____

_____

_____

_____

_____

_____

Notes

CHAPTER 10 Sketches, Mock-Ups, and Working Drawings

Chapter Review

Carefully read Chapter 10 of the text and answer the following questions.

1. Three-dimensional replicas of a design made of convenient, inexpensive materials are known as _____.

_____

2. _____ communicates ideas with a drawing.

_____

3. Identify the following types of sketches shown.

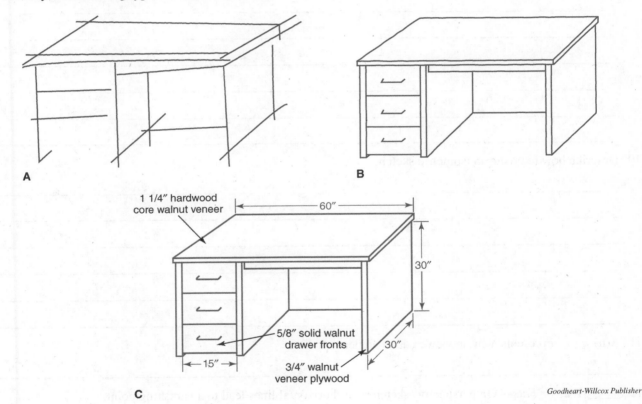

A

B

1 1/4" hardwood
core walnut veneer

60"

30"

5/8" solid walnut
drawer fronts

15"

3/4" walnut
veneer plywood

30"

C

Goodheart-Willcox Publisher

A. _____

B. _____

C. _____

4. _____ Which sketch includes most of the information to be put on a working drawing, such as dimensions and materials?
   A. Rough
   B. Refined
   C. Thumbnail
   D. Assembly

5. A(n) _____ sketch represents a product as seen from a corner.

_____

6. A(n) _____ sketch represents a product as if it were viewed from the front.

_____

7. _____ *True or False?* Isometric sketches represent a true view of an object.

8. List four skills necessary for successful sketching.

_____

_____

_____

_____

9. Describe how to create a cabinet oblique sketch.

_____

_____

_____

_____

_____

10. Describe how to create an isometric sketch.

_____

_____

_____

_____

_____

11. A(n) _____ represents a circle viewed at an angle.

_____

12. _____ *True or False?* On perspective sketches, all horizontal lines lead to a vanishing point.

13. _____ Working drawings include _____.
   A. views of the product showing style and features
   B. dimensions and a plan of procedure
   C. a list of materials and supplies
   D. All of the above.

14. How are architectural drawings used?

_____

_____

_____

_____

15. _____ Which of the following statements regarding floor plans is *true*?
   A. They explain where built-in cabinetry is to be located.
   B. They are used to determine the size and style of cabinets to be built.
   C. They represent the front view of built-in cabinetry.
   D. They indicate who should install the built-in cabinetry.

16. What information is found on material specifications?

   _____

   _____

   _____

17. Drawings submitted for approval prior to fabrication are known as _____ drawings.

   _____

18. List the logical order to follow when reading shop drawings.

   _____

   _____

   _____

   _____

Match each term with the correct description.

19. _____ Show the layout of the product as if it were unfolded.

20. _____ Allows you to see an object as if material were cut away.

21. _____ Represents a picture of the final product.

22. _____ Separate drawing of individual components or joints.

23. _____ Often found on pictorial drawings, it includes a separate list of parts.

24. _____ Shows the product disassembled.

25. _____ Rectangular space on each page of a set of drawings.

26. _____ Two or more views are shown to describe the product.

27. A(n) _____ mock-up looks like the final product but is not functional.

   _____

A. Pictorial view
B. Title block
C. Development drawing
D. Detail drawing
E. Multiview drawing
F. Parts balloon
G. Section drawing
H. Exploded view

 # Activity 10A

Bookshelf

Make a cabinet oblique sketch of a typical four-shelf bookcase of your own design on the graph paper provided.

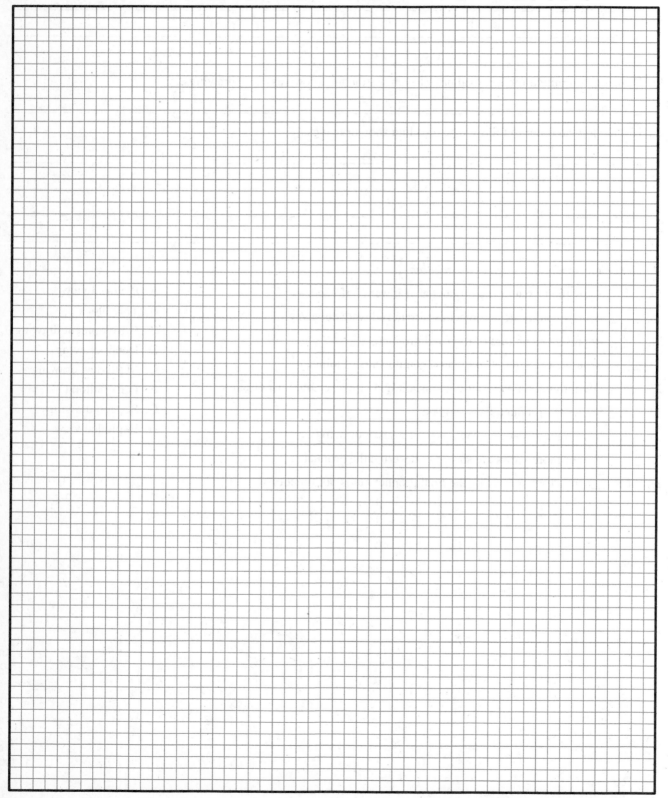

CHAPTER 11 Creating Working Drawings

Chapter Review

Carefully read Chapter 11 of the text and answer the following questions.

1. Producing working drawings for cabinets, case goods, chairs, and tables is part of _____ technology.

_____

2. Identify each of the following lines.

Goodheart-Willcox Publisher

A. _____

Goodheart-Willcox Publisher

B. _____

Goodheart-Willcox Publisher

C. _____

Goodheart-Willcox Publisher

D. _____

3. List the abbreviations used on architectural and shop drawings for each of the following terms.

Architectural Drawings

A. _____ —ceiling

B. _____ —flooring

C. _____ —between or on center

D. _____ —cabinet

E. _____ —closet

F. _____ —centerline

G. _____ —counter

H. _____ —minimum

I. _____ —maximum

J. _____ —length

K. _____ —height

L. _____ —dishwasher

M. _____ —refrigerator

N. _____ —vanity

Shop Drawings

A. _____ —radius   I. _____ —finish

B. _____ —diameter   J. _____ —laminate

C. _____ —assembly   K. _____ —maximum

D. _____ —average   L. _____ —minimum

E. _____ —dimension   M. _____ —number

F. _____ —drawing   N. _____ —round

G. _____ —inside diameter   O. _____ —with

H. _____ —outside diameter   P. _____ —without

Match each term with the correct description.

4. _____ Used to draw lines vertical to or at an angle to a T-square.

5. _____ Used for making circles and arcs.

6. _____ Used to draw straight lines.

7. _____ Moves up and down a drafting table and works much like a T-square.

8. _____ Used to transfer distances without marking the paper.

9. _____ Used to draw common shapes.

10. _____ High quality, erasable material used for drawing and copying.

11. _____ Used to control the amount of space covered by pictorial, multiview, and detail drawings.

12. _____ allow you to control the amount of space covered by pictorial, multiview, and detail drawings.

A. Straightedge

B. Polyester film

C. Triangle

D. Parallel rule

E. Scale

F. Template

G. Compass

H. Divider

_____

13. List the four essential parts of a shop drawing.

_____

_____

_____

_____

14. What types of information should appear on the title block?

_____

_____

_____

_____

15. When creating a(n) _____ drawing, the front view should be the face of the product with the most features.

_____

16. _____ includes both linear and radial distances using extension, dimension, leader, and radius dimension lines.

_____

17. Identify each type of dimensioning line.

A. _____

B. _____

C. _____

D. _____

E. _____

5 1/4

3/4R

1/2 R

3/4 DIA.

1/2 DRILL
3 HOLES

Goodheart-Willcox Publisher

18. _____ Lettering on shop drawings is _____.
   A. Gothic, lowercase letters
   B. Gothic, uppercase letters
   C. Gothic, both upper- and lowercase letters
   D. Times New Roman, uppercase letters

19. When drawing _____ of repeated objects, draw only one object completely and block in the others.

_____

20. List five items of information that are included on a bill of materials.

_____

_____

_____

_____

_____

21. _____ refers to quantities of materials and supplies as you buy them.

_____

22. List five functions computers can perform in relation to the cabinetmaking process.

_____

_____

_____

_____

_____

 Activity 11A

Alphabet of Lines

Using Figure 11-1 in the textbook, re-create and label each line in the alphabet of lines.

 Activity 11B

Lettering

Lettering is not printing or writing; it is more like drawing because specific strokes are made for each letter and number. In this activity, make five copies of each letter and numeral, using the strokes shown in Figure 11-13 in the textbook. Use the lines as the upper and lower construction lines for your letters. Skip a space between each row of lettering.

Notes

| CHAPTER 12 | Measuring, Marking, and Laying Out Materials |

Chapter Review

Carefully read Chapter 12 of the text and answer the following questions.

1. _____ Squareness means that all corners join at a(n) _____° angle.
   A. 30
   B. 45
   C. 90
   D. 180

2. Most cabinetmakers mark workpieces with _____.

   _____

3. What procedures are taking place in the following illustrations?

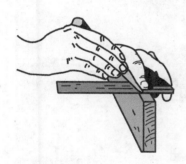

Stanley Tools

A. _____   B. _____

4. The _____ is designed to make parallel lines on stock.

   _____

5. Name the two systems followed by measuring tools.

   _____

   _____

6. _____ Which of the following will measure both straight lengths and curves?
   A. Flexible rule
   B. Depth gauge
   C. Caliper rule
   D. All of the above.

7. List the three purposes of squares.

   _____

   _____

   _____

Identify each piece of measuring equipment shown.

8. _____

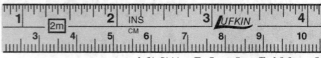

9. _____

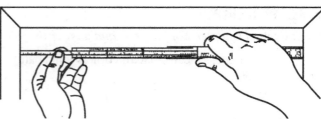

10. _____

11. _____

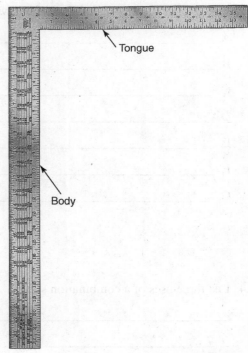

Tongue

Body

Stanley Tools

12. _____

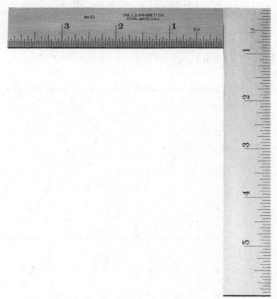

The L.S. Starrett Co.

13. Identify the parts indicated on the following combination square.

A. _____

B. _____

C. _____

D. _____

E. _____

F. _____

G. _____

The L.S. Starrett Co.

14. List three uses of a combination square.

_____

_____

_____

15. _____ tools transfer distances, angles, and contours.

_____

Identify each piece of layout equipment shown.

16. _____

Goodheart-Willcox Publisher

17. _____

General Hardware

Name _____

18. _____

19. _____

20. _____

21. _____

22. _____

23. _____

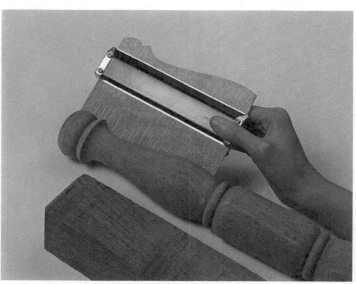

24. _____ When marking a distance on a piece of wood, use _____.
    A. an ink marker and make a dot
    B. a pencil and make an arrow
    C. chalk and make a square
    D. a crayon and make a circle

25. Most _____ are made using a rule or square.

    _____

26. _____ Which of the following can be used to make accurate circles and arcs?
    A. Compasses
    B. Dividers
    C. Trammel points
    D. All of the above.

27. Name two tools commonly used to lay out polygons.

    _____

    _____

28. _____ A _____ pattern is a way to transfer complex designs from working drawings to material.
    A. half
    B. detail
    C. square grid
    D. rectangle

29. A(n) _____ is a permanent full-size pattern used for guiding a tool.

    _____

30. _____ *True or False?* Moving joints on measuring and layout tools can be lubricated with paste wax.

Transferring a Design

Using the technique explained in Section 12.4.5 of the textbook, enlarge the design in Figure 12-34. Use the following grid. For this activity, the grid is marked out for you. You can make your own grid for future patterns using layout tools such as a framing square and heavy paper.

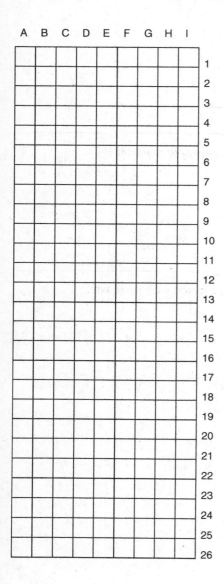

<table>
<tr><td>CHAPTER
13</td><td># Wood Characteristics</td></tr>
</table>

Chapter Review

Carefully read Chapter 13 of the text and answer the following questions.

1. List five desirable qualities of wood.

    _____

    _____

    _____

    _____

    _____

2. _____ *True or False?* Wood is an inelastic material.

3. Identify the parts of a tree.

    A. _____

    B. _____

    C. _____

    D. _____

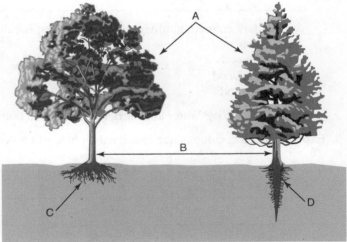

Goodheart-Willcox Publisher

4. Identify the layers of a tree.

A. _____

B. _____

C. _____

D. _____

E. _____

F. _____

G. _____

Forest Products Laboratory

5. _____ are created by the growth of a tree that occurs in a single growing season.

_____

6. _____ The dark colored, nonliving section of a tree is called _____.
   A. heartwood
   B. sapwood
   C. earlywood
   D. latewood

7. _____ *True or False?* Water and nutrients are carried outward from the center of a tree by wood rays.

8. Explain what a deciduous tree is and name three examples of it.

_____

_____

_____

_____

9. Explain what a coniferous tree is and name three examples of it.

_____

_____

_____

_____

10. Wood from conifers is called _____.

_____

11. Identify each type of wood face shown.

A. _____

B. _____

C. _____

A   B   C

Forest Products Laboratory

Match each term with the correct definition.

12. _____ Main passages for liquid moving from roots to crown.

13. _____ Vertical cells in hardwood.

14. _____ Move nutrients between the center and outer portions of a tree.

15. _____ Substance that holds cells together.

16. _____ Formed when space between cells expands.

17. _____ Very small cells used for additional food storage.

18. _____ Vertical cells that move liquids from the roots to the crown.

19. _____ _____-porous hardwoods have similar size pores throughout the growing
season.
A. Ring
B. Semiring
C. Diffuse
D. None of the above.

A. Parenchyma cells
B. Fibers
C. Tracheids
D. Resin ducts
E. Lignin
F. Rays
G. Vessels

20. Name four factors that determine the appearance of wood.

_____

_____

_____

_____

21. The pattern of lines visible in sawn lumber that is formed by the annual rings is called _____.

_____

22. _____ describes the amount of water in the wood cells.

_____

23. Identify the item being used in the following illustration.

_____

Patrick A. Molzahn

24. Provide the formula for and outline the procedure for calculating wood moisture content.

_____

_____

_____

_____

_____

_____

25. _____ *True or False?* Equilibrium moisture content does *not* change with the seasons.

26. Write the formula for calculating wood shrinkage.

_____

27. Wood _____ at different rates in different directions.

_____

28. List three factors that control the weight of wood.

_____

_____

_____

29. Write out the formula for calculating specific gravity.

_____

30. _____ *True or False?* Woods with high specific gravity tend to dull tools faster.

31. _____ wood is in leaning trees of some hardwood species.

_____

32. _____ is the capability of wood to spring back after being bent.

_____

 Activity 13A

Chart of Wood Characteristics

Create a chart to record the characteristics of the wood in your workshop. Using your textbook as a guide, create a list of characteristics that should be noted. In the first column, record the name of each type of wood. In following columns, list the characteristics of that wood. For example, column headings might include color, tree classification (deciduous or coniferous), and moisture content.

The objective of this activity is to create a useful chart that you can fill out as you progress through your course and as you encounter different types of wood. Create your chart in the space provided.

CHAPTER 14 / Lumber and Millwork

Chapter Review

Carefully read Chapter 14 of the text and answer the following questions.

1. List the sequence of steps used to bring wood to market as lumber.

   _____

   _____

   _____

   _____

2. _____ felling is the harvesting of large sections of forest at one time using heavy machinery.

   _____

3. _____ felling is the harvesting of single trees.

   _____

4. _____ What is the most common method of sawing?
   A. Plain sawing
   B. Quarter sawing
   C. Rift sawing
   D. Ripping

5. _____ *True or False?* Quarter-sawn lumber twists and cups more than plain-sawn lumber.

6. The process of drying lumber is known as _____.

   _____

7. _____ Wood used for cabinetmaking should be between _____ percent moisture content.
   A. 2 and 4
   B. 6 and 8
   C. 10 and 12
   D. 15 and 19

8. What process is shown in the following illustration?

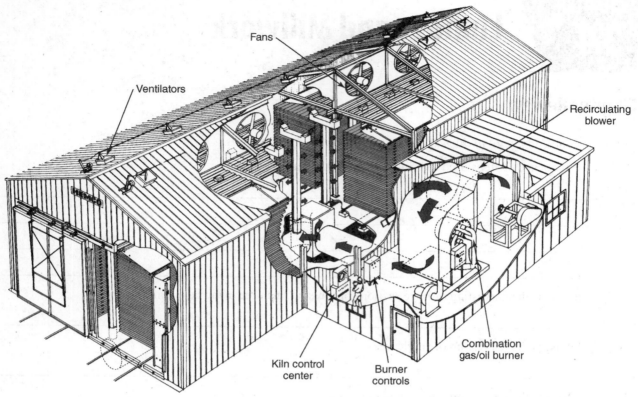

Fans

Ventilators

Recirculating
blower

Kiln control
center

Burner
controls

Combination
gas/oil burner

Western Wood Products Assoc.; Harvey Engineering and Manufacturing Corp.

_____

9. Name three categories of lumber defects.

_____

_____

_____

10. List four types of natural defects in wood.

_____

_____

_____

_____

11. _____ *True or False?* Pitch pockets are openings in wood that contain solid or liquid resins.

12. _____ Peck is caused by _____.
    A. insects
    B. fungus
    C. moisture
    D. dryness

13. List eight types of defects caused by improper seasoning or storage.

_____

_____

_____

_____

_____

_____

_____

_____

14. _____ Which of the following types of warp forms a curve lengthwise along the face of a board from end to end?
    A. Bow
    B. Crook
    C. Cup
    D. Twist

15. _____ is the disintegration of wood fibers due to fungi.

_____

16. _____ Wood preservatives can be used to prevent _____.
    A. stain
    B. decay
    C. insect damage
    D. All of the above.

17. Machining defects occur most often during _____.

_____

18. _____ Which of the following occurs when boards are fed into the surfacer faster than the knives can cut?
    A. Torn grain
    B. Raised grain
    C. Wavy dressing
    D. Skip

19. _____ Which of the following can cause machine burn?
    A. Dull tools
    B. Too slow of feed
    C. The cutter head rubbing in one place if the board stops during surfacing
    D. All of the above.

20. How is lumber rated by grades?

_____

_____

_____

21. _____ grades are based on the amount of clear lumber that can be cut from a board.

_____

22. Name three types of construction-grade lumber.

_____

_____

_____

23. _____-grade lumber is more suited to cabinetmaking than construction-grade lumber.

_____

24. _____ removes 1/8″–1/4″ (3 mm–6 mm) from the nominal (rough) size.

_____

25. Why would a cabinetmaker specify KD lumber on an order form?

_____

_____

26. _____ consists of specialty items frequently processed from moulding-grade lumber.

_____

27. _____ _____-grade wood mouldings are suitable for natural or clear finishes.
    A. R
    B. P
    C. N
    D. M

28. When ordering millwork, what should be specified?

_____

_____

_____

29. _____ are used as both support and decoration on stair rails.

_____

30. Identify the following plugs.

    A. _____

    B. _____

    C. _____

A   B ←Diam→   C

Goodheart-Willcox Publisher

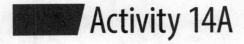

Activity 14A

Specialty Item

Brainstorm ideas for a specialty item that could be mass-produced in your class and sold to other cabinetmaking businesses as an add-on for their cabinets. Write your ideas in the space provided. See Figure 14-34 in the textbook for inspiration and ideas.

Notes

CHAPTER 15

Cabinet and Furniture Woods

Chapter Review

Carefully read Chapter 15 of the text and answer the following questions.

Match each term with the correct description.

1. _____ Botanical name of a tree, given in Latin.

2. _____ Wood cut from coniferous trees.

3. _____ Growth that occurs in the spring.

4. _____ Growth that occurs in the summer.

5. _____ A ring caused by the addition of earlywood and latewood growth to the trunk of a tree.

6. _____ Wormholes are not a defect.

7. _____ Wood that has large pores that are cut open during machining.

8. _____ Wood that has small pores.

9. _____ Growth that occurs across the normal grain direction.

10. _____ The smooth or rough feel of the wood surface.

11. _____ A measure of how likely wood is to swell or shrink when exposed to moisture.

12. _____ A shiny appearance when sanded smooth.

13. _____ The outer layers of a tree that carry nutrients and water.

14. _____ Tells how easily the wood can be cut, surfaced, sanded, or processed by other means.

15. _____ Wormholes are a defect.

16. _____ A measure of the mass of wood.

17. _____ Wood cut from deciduous trees.

18. _____ Classification of trees.

19. _____ The inner layers of a tree that consist of inactive cells.

A. Texture

B. Luster

C. Open grain

D. Density

E. Latewood

F. Sapwood

G. Annual ring

H. Hardwood

I. Dimensional stability

J. Heartwood

K. Softwood

L. Closed grain

M. Genus

N. Species

O. Machining/working qualities

P. Cross grain

Q. Earlywood

R. WHAD

S. WHND

20. Various working properties of woods are provided. Next to each property, list at least one species of wood that you would choose that exhibits that property. Use the "Summary of Wood Characteristics" chart in the textbook's Appendix for reference.

Planing—excellent _____

Drilling—excellent _____

Sanding—excellent _____

Turning—excellent _____

Gluing—excellent _____

Nail and screw holding—excellent _____

Bending—difficult _____

Bending—easy _____

Hardness—hard _____

Hardness—soft _____

Compression strength—weak _____

Compression strength—strong _____

Shock resistance—low _____

Shock resistance—high _____

Stiffness—limber _____

Stiffness—stiff _____

Availability—lumber _____

Availability—veneer _____

Cost—expensive _____

Cost—inexpensive _____

21. List five hardwoods that require filler to make a smooth surface for finishing.

_____

_____

_____

_____

_____

 Activity 15A

Wood Characteristics

Provide a brief description of the characteristics of each wood and list two of its uses.

1. Alder, red: _____

_____

_____

2. Ash: _____

_____

_____

3. Banak: _____

_____

_____

4. Basswood: _____

_____

_____

5. Beech: _____

_____

_____

6. Birch: _____

_____

_____

7. Butternut: _____

_____

_____

8. Cedar, aromatic red: _____

_____

_____

9. Cherry: _____

_____

_____

10. Chestnut: _____

_____

_____

11. Cottonwood: _____

_____

_____

12. Cypress, bald: _____

_____

_____

13. Ebony: _____

_____

_____

14. Elm, American: _____

_____

_____

15. Fir, Douglas: _____

_____

_____

16. Gum, red: _____

_____

_____

17. Hackberry: _____

_____

_____

18. Hickory: _____

_____

_____

19. Lauan: _____

_____

_____

20. Limba: _____

_____

_____

Name _____

21. Mahoganies, genuine: _____

_____

_____

22. Mahogany, African: _____

_____

_____

23. Maple, hard: _____

_____

_____

24. Maple, soft: _____

_____

_____

25. Oak, red or white: _____

_____

_____

26. Paldao: _____

_____

_____

27. Pecan: _____

_____

_____

28. Pine, Ponderosa: _____

_____

_____

29. Pine, sugar: _____

_____

_____

30. Pine, yellow: _____

_____

_____

31. Primavera: _____

_____

_____

32. Redwood: _____

_____

_____

33. Rosewood: _____

_____

_____

34. Santos Rosewood: _____

_____

_____

35. Sapele: _____

_____

_____

36. Sassafras: _____

_____

_____

37. Satinwood: _____

_____

_____

38. Spruce: _____

_____

_____

39. Sycamore: _____

_____

_____

40. Teak: _____

_____

_____

41. Tulip, American (Yellow Poplar): _____

_____

_____

42. Walnut, American: _____

_____

_____

43. Willow: _____

_____

_____

44. Zebrawood: _____

_____

_____

Activity 15B

What Wood to Use?

Imagine you are in the business of making rocking chairs. What wood would you use? Explain how you came to your decision and why you picked the wood you did. Include considerations such as cost, appearance, and workability, along with other design decisions.

_____

_____

_____

_____

_____

_____

_____

_____

_____

_____

_____

_____

_____

_____

_____

_____

_____

_____

_____

_____

_____

_____

CHAPTER 16 — Manufactured Panel Products

Chapter Review

Carefully read Chapter 16 of the text and answer the following questions.

1. _____ Manufactured panel products are widely used by cabinetmakers to _____.
   A. create large surfaces for case goods
   B. reduce the need for edge gluing lumber to make wide boards
   C. reduce production time without sacrificing quality
   D. All of the above.

2. _____ *True or False?* Panel products are typically less stable than solid lumber.

3. List three categories of panel products.

   _____

   _____

   _____

4. _____ Structural wood panels are selected when _____ are required.
   A. beauty and appearance
   B. stability and strength
   C. strength and appearance
   D. flexibility and structure

5. What structural panel is manufactured with a core material sandwiched between two pieces of face veneer?

   _____

6. The thickness of veneer-core plywood refers to the number of _____.

   _____

7. Identify the following type of plywood and identify each item indicated on the diagram.

   _____

Goodheart-Willcox Publisher

A. _____   C. _____

B. _____   D. _____

8. _____ Which of the following statements regarding lumber-core plywood is *false*?
   A. It has a solid wood center and thin veneer faces.
   B. The core may be thin, laminated strips of wood or wider boards.
   C. The veneer layers between the core and both faces are called cross-bands.
   D. The grain of cross-bands is at 45° to the faces.

9. _____ Particleboard-core plywood is _____.
   A. more expensive than veneer-core plywood
   B. similar in price to veneer-core plywood
   C. more expensive than lumber-core plywood
   D. approximately the same price as lumber-core plywood

10. Identify the items indicated on the grade trademark.

   A. _____

   B. _____

   C. _____

   D. _____

   E. _____

   F. _____

   G. _____

   H. _____

TYPICAL APA SANDED PLYWOOD TRADEMARKS

Typical Back Stamp

APA

A —— **A-C** GROUP 1 —— E
B ——
C —— EXTERIOR
THICKNESS 0.453 IN.
D —— 000 —— F
PS 1-09
15/32 CATEGORY —— H —— G

APA—The Engineered Wood Association

11. The _____ that bonds the layers determines, in part, how the plywood is used.

   _____

Match each term with the correct description.

12. _____ Made of wood chip or fiber core faced with a veneer.

13. _____ Made of wood wafers and resin adhesive.

14. _____ Composed of small wood flakes, chips, and shavings bonded together with resins or adhesives.

15. _____ Manufactured with strands of wood that are layered perpendicular to each other.

16. _____ Designed to span specified distances.

17. _____ Adhesive designed to have resistance to water, acid, and resin solvents.

18. _____ panels provide the appearance and strength of solid hardwood, yet they are much less expensive.

A. Oriented strand board

B. Composite panel

C. Structural particleboard

D. Performance-rated structural wood panels

E. Waferboard

F. Phenolic resin

19. List four items that determine the appearance of face veneers of hardwood plywood.

   _____

   _____

   _____

   _____

20. Standards for hardwood plywood are set by the _____.

   _____

21. List ten specifications to include when ordering hardwood plywood.

_____

_____

_____

_____

_____

_____

_____

_____

_____

_____

22. _____ *True or False?* Low-density fiberboard and particleboard are manufactured using heat and pressure.

23. _____ Which of the following types of hardboard is the strongest and has good finishing qualities?
    A. Standard hardboard
    B. Tempered hardboard
    C. Service hardboard
    D. Plywood hardboard

24. Medium-density fiberboard (MDF) is _____ than hardboard.

_____

25. Because of its smooth surface, _____ is often used as substrate material for laminations on cabinets, countertops, tabletops, and drawer fronts.

_____

26. _____ particleboard is a low-cost alternative to fiberboard.

_____

27. The nail- or screw-holding ability of a panel is related to its _____.

_____

 **Activity 16A**

Choosing the Right Material for the Job

Imagine you are the owner of a cabinet shop and you want to build the best cabinets for the least amount of money. Based on your study of Chapter 16, explain what materials decisions you would make for the following.

1. Case parts that will be seen: _____

_____

_____

_____

_____

_____

_____

_____

2. Case parts that will not be seen: _____

_____

_____

_____

_____

_____

_____

_____

3. Countertops that will be laminated: _____

_____

_____

_____

_____

_____

_____

_____

CHAPTER 17

Veneers and Plastic Overlays

Chapter Review

Carefully read Chapter 17 of the text and answer the following questions.

1. A(n) _____ is any thin sheet material that covers a core material.

   _____

2. Name the two types of veneer.

   _____

   _____

3. _____ The most common thickness of _____ veneer is 1/42″.
   A. flat
   B. flexible
   C. overlay
   D. substrate

4. Logs from which veneer is cut are debarked and cut to length to form _____.

   _____

5. The appearance of veneer depends greatly on the _____.

   _____

6. _____ Swinging the flitch against the knife is known as _____.
   A. rotary cutting
   B. slicing
   C. stay-log cutting
   D. chuck turning

7. Identify each of the following methods of cutting veneer.

   A. _____

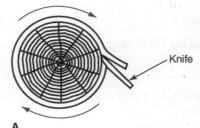

Goodheart-Willcox Publisher

   B. _____

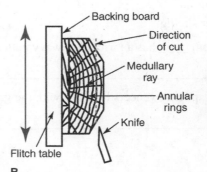

Goodheart-Willcox Publisher

C. _____

Direction of cut
Medullary ray
Annular rings
Knife

Flitch table
C

Goodheart-Willcox Publisher

D. _____

Direction of cut
Medullary ray
Annular rings
Knife
Center of lathe
Rotation of flitch
D

Goodheart-Willcox Publisher

E. _____

Annular rings
Direction of cut
Medullary ray
Knife
Rotation of flitch
E

Goodheart-Willcox Publisher

8. _____ The grain pattern created by most _____ is wide.
   A. rotary cutting
   B. flat slicing
   C. quarter-slicing
   D. rift cutting

9. Logs are sectioned into four flitches for _____.

_____

10. _____ _____ is a method of stay-log cutting that produces a large, U-patterned grain.
   A. Flat slicing
   B. Half-round cutting
   C. Rift cutting
   D. Quarter slicing

11. Veneer sliced from _____ has an irregular, wavy pattern.

_____

12. _____ is done by splicing veneers together with the grain pattern in specific directions.

_____

Name _____

13. Describe how each veneer match is made.

Book match: _____
_____

Slip match: _____
_____

Diamond match: _____
_____

Reverse-diamond match: _____
_____

Four-way center and butt match: _____
_____

Vertical butt and horizontal book leaf match: _____
_____

14. Identify each type of veneer matching.

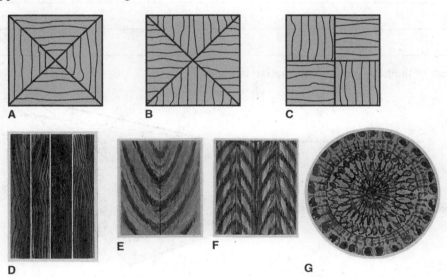

A. _____   E. _____

B. _____   F. _____

C. _____   G. _____

D. _____

15. Veneer _____ are made by cutting veneer into a pattern and bonding it to a wood backing.

_____

16. What is being done in the following photo?

_____

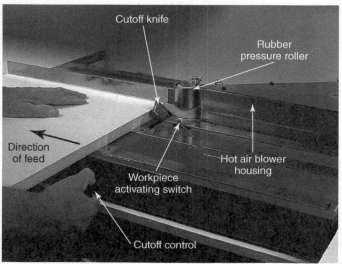

17. _____ *True or False?* Plastic overlays can be either rigid or flexible.

18. List three uses for plastic overlays.

_____

_____

_____

19. Identify the layers of the high-pressure decorative laminate.

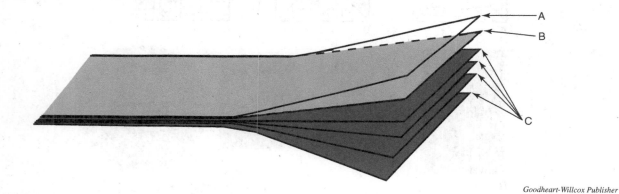

A. _____

B. _____

C. _____

Match each type of HPDL to the correct use.

20. _____ Provides a decorative surface in areas where there will be little wear.

21. _____ Useful for areas that have curved corners and edges.

22. _____ Applied to the opposite of a substrate covered by a decorative laminate.

23. _____ Most widely used type of laminate.

24. _____ Greater abrasion and scuff resistance than conventional laminate; used in commercial, contract, and institutional settings.

25. _____ Designed for use as wall panels and cabinet surfaces, especially where thinner and lighter material is needed.

26. _____ Very thick HPDL made of layers of phenolic resin-saturated kraft paper.

27. _____ *True or False?* The major use of LPDL panels is for cabinet exteriors.

A. General-purpose
B. VGL and VGS
C. Post-forming
D. Cabinet-liner
E. Backing sheet
F. High-wear
G. Compact laminate

28. Identify the parts of this flexible overlay.

A. _____

B. _____

Goodheart-Willcox Publisher

Activity 17A

Material Warpage

Look at the illustration provided. It is a cross section of a piece of industrial particleboard. Laminate is glued to one side, with no backer board glued to the other side. Draw a similar illustration showing how the piece will warp if placed in a humid environment. Remember that wood expands when it takes in moisture.

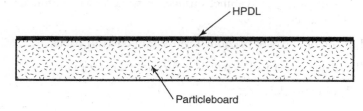

HPDL

Particleboard

CHAPTER 18 / Glass, Plastic, and Related Products

Chapter Review

Carefully read Chapter 18 of the text and answer the following questions.

Match each term with the correct description.

1. _____ Ensures a controlled break when cutting glass.

2. _____ Distort beyond use if reheated.

3. _____ Can be reheated and reformed many times.

4. _____ Process in which glass is reheated and quenched quickly to increase strength.

5. _____ Synthetic compounds, also called resins.

6. _____ Made of silica, soda ash, and limestone.

A. Thermoplastic materials
B. Annealing
C. Glass
D. Thermoset plastics
E. Tempered
F. Plastics

7. Name three forms glass and plastic products can take.

_____

_____

_____

8. List three uses of flat glass in cabinetry.

_____

_____

_____

9. _____ glass is thicker and often stronger than sheet glass.

_____

10. _____ Glass that has been worked in some way to manipulate the surface, color, or pattern of the glass is called _____ glass.
   A. flat
   B. float
   C. tinted
   D. decorative

11. _____ glass is made by adding coloring agents to molten glass.

_____

12. _____ *True or False?* Glass is installed before a cabinet has been assembled and finished.

13. _____ Which of the following statements regarding glass cutting is *false*?
   A. The glass is actually cut.
   B. It can be done by hand or machine.
   C. The cutting wheel can be pushed or pulled across the glass surface.
   D. Lubricant is used to reduce the amount of glass-surface flaking.

14. Glass has a tendency to _____, making it very difficult to break cleanly.

_____

15. _____ Glass can be fractured by _____.
   A. bending glass clamped to a scoring machine
   B. bending with pliers
   C. tapping
   D. All of the above.

16. List three methods for mounting glass.

_____

_____

_____

17. _____ glass refers to panels using colorless glass.

_____

18. Place the following steps for preparing and assembling leaded and stained glass panels in order with the appropriate step number (1–6).
   A. _____ Mounting the glass in the frame
   B. _____ Laying out full-size patterns
   C. _____ Fitting lead came
   D. _____ Cutting glass
   E. _____ Grouting along the lead seams
   F. _____ Soldering the lead joints

19. _____ plastic is a rigid plastic often used to replace glass in many applications.

_____

20. Liquid _____ resin is used with fiberglass and as a casting and coating material.

_____

21. _____ Which of the following can be either a thermoplastic or thermoset plastic?
   A. Polystyrene
   B. Polyethylene
   C. Polyurethane
   D. None of the above.

22. Name two ways plastic can be cut.

_____

_____

23. Why is a backer board placed under plastic when it is being drilled?

_____

_____

24. _____ *True or False?* When working with plastics, finishing should be necessary only on the sheet edges.

25. _____ dissolve plastic, causing joint surfaces to soften and the plastic to flow together.

_____

26. Identify the decorative edge designs shown.

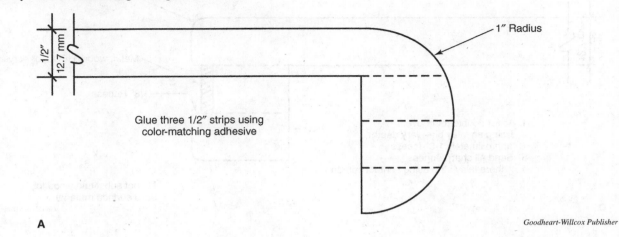

Glue three 1/2″ strips using color-matching adhesive

1″ Radius

1/2″ 12.7 mm

Goodheart-Willcox Publisher

A

A. _____

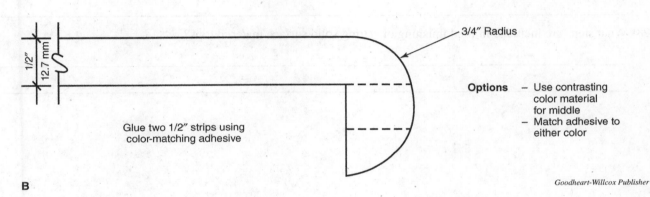

Glue two 1/2″ strips using color-matching adhesive

3/4″ Radius

1/2″ 12.7 mm

Options – Use contrasting color material for middle
– Match adhesive to either color

B

Goodheart-Willcox Publisher

B. _____

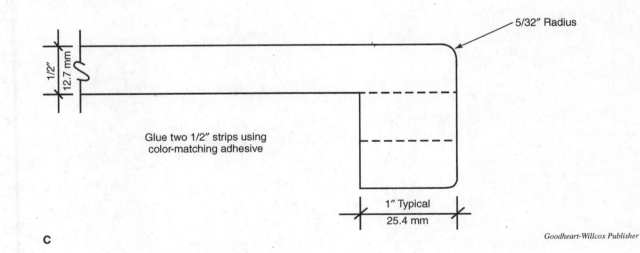

Glue two 1/2″ strips using color-matching adhesive

5/32″ Radius

1/2″ 12.7 mm

1″ Typical 25.4 mm

C

Goodheart-Willcox Publisher

C. _____

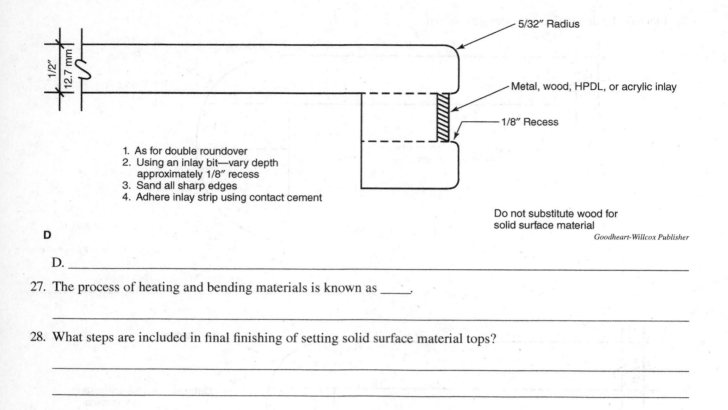

1. As for double roundover
2. Using an inlay bit—vary depth approximately 1/8″ recess
3. Sand all sharp edges
4. Adhere inlay strip using contact cement

5/32″ Radius

Metal, wood, HPDL, or acrylic inlay

1/8″ Recess

Do not substitute wood for solid surface material

Goodheart-Willcox Publisher

D

D. _____

27. The process of heating and bending materials is known as _____.

_____

28. What steps are included in final finishing of setting solid surface material tops?

_____

_____

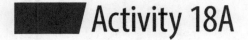

Activity 18A

Countertop Materials

You have been contracted to build a countertop for a convenience store. This countertop must withstand constant use. Draw a cross-section diagram and specify the materials you would use for the countertop. Evaluation of this activity will be based on neatness, accuracy, scale, and drawing technique. It is essential that the cabinetmaker become skilled at making scaled drawings and sketches as outlined in Chapter 10 of the textbook. Make sure you use a sharp pencil when sketching. Your instructor may give you the option to use the computer or a CAD program to create the diagram.

Notes

CHAPTER 19 / Hardware

Chapter Review

Carefully read Chapter 19 of the text and answer the following questions.

1. _____ Which of the following types of pulls include a backplate that supports a hinged pull?
   A. Surface mount
   B. Flush mount
   C. Bail
   D. Ring

2. _____ *True or False?* Upper and lower sliding door tracks must be perpendicular.

3. Identify the following case front styles shown.

Corner view (Countertop removed)	Top view	Corner view (Countertop removed)	Top view	Corner view (Countertop removed)	Top view
A		B		C	
D		E		F	

Goodheart-Willcox Publisher

A. _____   D. _____

B. _____   E. _____

C. _____   F. _____

4. When using three or more hinges, the pins must be in line, otherwise the door may become _____.

_____

5. Name and describe two types of butt hinges.

_____

_____

_____

_____

_____

6. Identify each type of hinge shown and name at least one application for it.

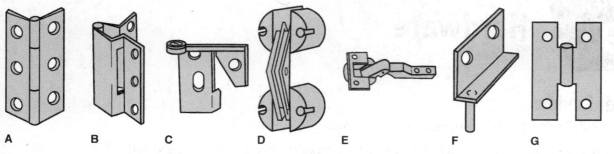

A   B   C   D   E   F   G

Goodheart-Willcox Publisher

A. _____   E. _____

B. _____   F. _____

C. _____   G. _____

D. _____

7. Identify the components of the following European concealed hinge shown.

A. _____

B. _____

C. _____

D. _____

E. _____

F. _____

G. _____

H. _____

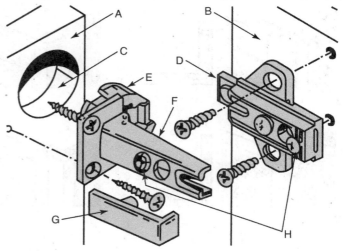

Häfele America Co.

8. Name three glass hinge styles.

_____

_____

_____

Match each term with the correct description.

9. _____ Consists of two spring-loaded balls with a strike between them.

10. _____ Consists of a bent spring steel catch and a ball head screw used as a strike.

11. _____ Hooks onto the strike and a spring-action lever is pushed to release the catch.

12. _____ Includes a spring arm with a roller that seats in a concave strike when the door is closed.

13. _____ Has one roller and a hook-shaped strike.

14. _____ Consists of a spring-operated catch that mounts into a hole drilled in the edge of the cabinet face or in the door.

15. _____ Has a permanent magnet that attracts a flat, plated steel strike.

16. Doors with latches do *not* require _____ since the hardware opens the door as well as holds it closed.

A. Friction catch
B. Single roller catch
C. Bullet catch
D. Elbow catch
E. Ball catch
F. Magnetic catch
G. Spring catch

_____

17. _____ _____ slides permit the entire drawer body to extend out of the drawer.
   A. Standard
   B. Full-extension
   C. Full-extension with over travel
   D. Half-extension

18. _____ *True or False?* A drawer track is attached to the inside of a cabinet.

19. _____ The body of a(n) _____-action lock is flat on one or both sides.
   A. bolt
   B. ratchet
   C. cam
   D. strip

20. Casters are mounted on either cabinet bottoms or _____ for mobility.

_____

21. _____ protect the bottoms and legs of freestanding cabinetry.

_____

 ## Activity 19A

Which Way to Open?

Imagine you own a cabinet shop. You wish to standardize the door style used on cabinets that are produced in your shop. Refer to Figure 19-10 in the textbook. Draw the door style you chose in the space provided and explain why you chose it. Evaluation of this activity will be based on neatness, accuracy, scale, and drawing technique. It is essential that the cabinetmaker become skilled at making scaled drawings and sketches as outlined in Chapter 10 of the textbook. Make sure you use a sharp pencil when sketching. Your instructor may give you the option to use the computer or a CAD program to create the diagram.

1. Why did you choose this door style?

_____

_____

_____

_____

CHAPTER 20 / Fasteners

Chapter Review

Carefully read Chapter 20 of the text and answer the following questions.

1. List four examples of fasteners.

   _____

   _____

   _____

   _____

2. _____ fasteners allow for easy assembly and disassembly.

   _____

3. _____ *True or False?* A nail has more holding power than a screw.

4. Identify each type of nail shown.

   A. _____

   B. _____

   C. _____

   D. _____

   E. _____

   F. _____

   Continental Steel Corp.

5. Most cabinetmakers drive nails using a(n) _____ fastening tool.

   _____

6. Nails are measured by _____ and specific dimensions.

   _____

7. _____ The length of a nail should be _____ times the thickness of the materials being fastened.
   A. two
   B. three
   C. four
   D. five

8. Drilling a(n) _____ hole will prevent a board from splitting when driving nails.

   _____

9. What fastening method is shown in the following drawing?

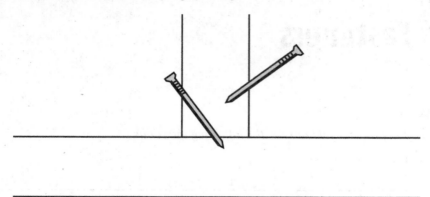

Goodheart-Willcox Publisher

_____

10. Which staple is sized correctly?

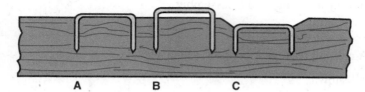

Goodheart-Willcox Publisher

_____

Match each term with the correct description.

11. _____ Eight-prong staples that join parts without cutting or splintering wood fibers.

12. _____ Small brass nails with round heads, used for decorative purposes.

13. _____ Angled fasteners.

14. _____ Installed in hidden areas, specified by crown width and leg length.

15. _____ Shaped pieces of steel.

16. _____ Sheet metal fasteners used on miter or butt joints in softwoods.

17. _____ Used primarily in upholstery.

18. A(n) _____ is a raised, helical rib or ridge around the interior or exterior of a cylindrically shaped fastener.

A. Skotch fasteners
B. Staples
C. Chevrons
D. Escutcheon pins
E. Tacks
F. Clamp nails
G. Corrugated fasteners

_____

Name _____

19. Identify each type of screw shown.

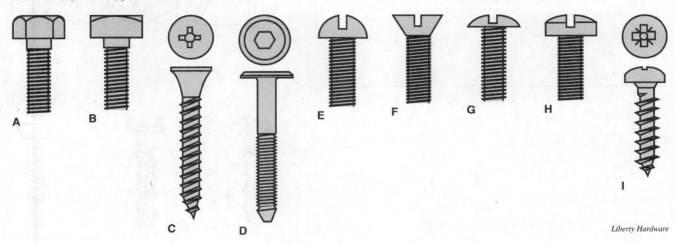

Liberty Hardware

A. _____   F. _____

B. _____   G. _____

C. _____   H. _____

D. _____   I. _____

E. _____

20. List the six steps for installing wood screws.

_____

_____

_____

_____

_____

_____

21. _____ bits may be used to drill pilot and clearance holes at the same time.

_____

22. Flat head screws and others with tapered heads require _____.

_____

23. Identify each type of screw shown.

A. _____

A

Liberty Hardware

B. _____

B

Liberty Hardware

C. _____

C

Liberty Hardware

D. _____

D

Goodheart-Willcox Publisher

E. _____

E

Graves-Humphreys, Inc.

F. _____

F

Graves-Humphreys, Inc.

24. Install _____ nuts when frequent assembly and disassembly is desired.

_____

25. Name three types of anchors.

_____

_____

_____

26. What is the purpose of repair plates?

_____

_____

27. _____ _____ connectors install quickly and can be used where appearance is not a factor.
   A. Bolt and cam
   B. Concave bolt
   C. Wedge pin
   D. Plug and socket

28. _____ _____ connectors are designed for joints that require less holding power.
   A. Bolt and cam
   B. Concave bolt
   C. Wedge pin
   D. Plug and socket

29. _____ Keku fasteners are used for installing _____.
   A. wainscoting
   B. framed mirrors
   C. wall panels
   D. All of the above.

Name ______________________ Date ____________ Class ________

Activity 20A

Purchasing Agent

Most manufacturing companies have purchasing agents. These individuals are responsible for finding materials and buying those materials at a cost that keeps the company competitive.

Imagine you are the purchasing agent for a cabinetmaking company. List several fasteners that you would buy in large quantities because they are used in all the cabinets your company produces. Explain why you included those fasteners in your inventory.

CHAPTER 21 / Ordering Materials and Supplies

Chapter Review

Carefully read Chapter 21 of the text and answer the following questions.

1. Being _____ means getting the right price for the right quantity at the right time at the quality level you need.

   _____

2. _____ Being thorough means you must study the _____ in order to create a complete, detailed, accurate order.
   A. product views
   B. bill of materials
   C. plan of procedure
   D. All of the above.

3. Evaluating the cost of goods from various vendors is known as _____ shopping.

   _____

4. _____ *True or False?* Liquids in larger containers are generally less expensive per ounce than those in smaller containers.

5. _____ are items that become part of the finished product.

   _____

Match each term with the correct description.

6. _____ Ordered by species and specific size.

7. _____ Usually sold in random widths and lengths.

8. _____ Ordered by shape or pattern number and linear foot.

9. _____ Ordered by the sheet, usually 4′ × 8′ (1220 mm × 2440 mm).

10. _____ Ordered by quantity, part number, manufacturer, finish, and size.

11. _____ Ordered by grit, type, and form.

    A. Abrasives
    B. Hardwood
    C. Softwood
    D. Fasteners
    E. Plywood
    F. Mouldings

12. _____ Glass is sized by _____.
   A. thickness
   B. dimensions
   C. any special treatments
   D. All of the above.

13. Name five specifications used when ordering fasteners.

   _____

   _____

   _____

   _____

14. _____ products include fillers, sealers, stains, and topcoatings.

   _____

15. Finishes require a compatible type of _____.

_____

16. _____ for cabinetmaking include consumable items that do *not* become a major part of the cabinet, such as abrasives, adhesives, rags, and wax.

_____

17. _____ _____ of abrasives include sheets, belts, disks, sleeves, pads, or powders.
   A. Grits
   B. Types
   C. Forms
   D. None of the above.

18. Check an adhesive's _____ before buying large quantities.

_____

19. _____ Stationary machinery should be purchased _____.
   A. by mail
   B. by phone
   C. at a store
   D. None of the above.

 Activity 21A

Material and Supplies Order Form

Companies often design their own business forms. A material order form is a typical form that a company would design. In this activity, you will make a material order form. Create your form on this page. Fill in the form with the supplies and materials listed in Figure 21-1. Leave space at the bottom of your form for totals. Figure costs using existing supply catalogues, the internet, or by calling suppliers. A personal visit to a supplier is the least efficient method, but may be necessary to see first-hand the quality of a product.

Notes

| CHAPTER 22 | Sawing with Hand and Portable Power Tools |

Chapter Review

Carefully read Chapter 22 of the text and answer the following questions.

1. Name the two groups of handsaws.

   _____

2. _____ Handsaws should be used to cut _____.
   A. high-density fiberboard
   B. particleboard
   C. plywood
   D. All of the above.

3. List four handsaws typically used for sawing straight lines.

   _____

   _____

   _____

   _____

4. A(n) _____ saw rips on one edge and crosscuts on the other.

   _____

5. What process is being done in the following photo?

   _____

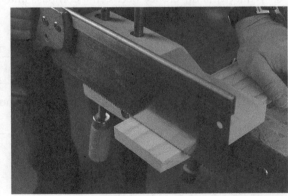

Patrick A. Molzahn

6. Identify the saw in the following drawing.

_____

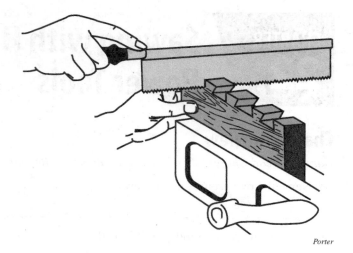

Porter

7. Identify the saw in the following photo.

_____

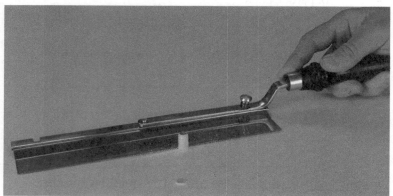

Patrick A. Molzahn

8. What is the difference between a compass and a keyhole saw?

_____

_____

_____

9. What type of saw is shown in the following photo?

_____

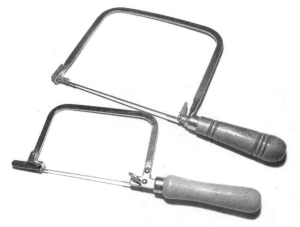

Jeff Banke/Shutterstock.com

10. _____ *True or False?* Select a narrow blade saw for straight workpieces and a wide blade saw for curved workpieces.

11. Use a(n) _____ motion when sawing by hand.

_____

12. List the steps in a proper hand sawing procedure.

_____

_____

_____

_____

_____

_____

_____

_____

13. Name four ways to support material before and after it is cut with a portable power saw.

_____

_____

_____

_____

14. The _____ saw is excellent for cutting lumber and panel products to approximate size.

_____

15. Name two adjustments on the portable circular saw.

_____

_____

16. To prevent tearout when cutting paneling or plywood with a circular saw, _____ the stock with a razor knife on the side of the blade opposite the offcut.

_____

17. Track saws have _____ on both sides of the blade to ensure that the cut has virtually no tearout.

_____

18. _____ cuts are sometimes made when cutting slots or pockets in material.

_____

19. _____ _____ saws are frequently used to cut stiles and rails to length.
    A. Combination
    B. Power miter
    C. Track
    D. Reciprocating

20. Identify the saw in the following photo.

_____

21. When sawing curved lines, make _____ in the offcut along the curve.

_____

22. Identify the saw in the following photo.

_____

23. _____ Proper maintenance of saw blades includes keeping them free of _____.
    A. moisture
    B. rust
    C. resin
    D. All of the above.

24. List four defects to check for when inspecting saw blades.

_____

_____

_____

_____

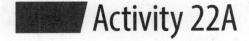

Activity 22A

Which Tool to Use?

Study the side view of the desk in the following drawing, and then respond to the following statements.

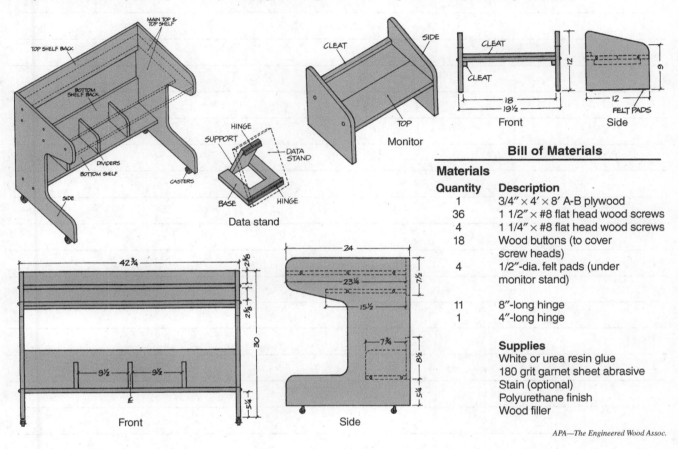

Bill of Materials

Materials

Quantity	Description
1	3/4″ × 4′ × 8′ A-B plywood
36	1 1/2″ × #8 flat head wood screws
4	1 1/4″ × #8 flat head wood screws
18	Wood buttons (to cover screw heads)
4	1/2″-dia. felt pads (under monitor stand)
11	8″-long hinge
1	4″-long hinge

Supplies
White or urea resin glue
180 grit garnet sheet abrasive
Stain (optional)
Polyurethane finish
Wood filler

APA—The Engineered Wood Assoc.

1. If you had to cut this part using hand tools, explain what tools you would use and the procedure you would follow.

   _____
   _____
   _____
   _____

2. If you had to cut this part using portable power tools, explain what tools you would use and the procedure you would follow.

   _____
   _____
   _____
   _____

Notes

CHAPTER 23 Sawing with Stationary Power Machines

Chapter Review

Carefully read Chapter 23 of the text and answer the following questions.

1. Stationary power-sawing machines are designed for what type of cuts?

   _____

   _____

2. Name four items to consider when selecting the proper saw.

   _____

   _____

   _____

   _____

3. List six safety tips to follow for safe and efficient operation of a power saw.

   _____

   _____

   _____

   _____

   _____

   _____

   _____

4. The most accurate straight-line sawing is done on equipment having a(n) _____ blade.

   _____

5. When sawing straight lines, the material must be _____ before and after the cut.

   _____

6. Identify the parts of the following table saw.

A. _____   E. _____

B. _____   F. _____

C. _____   G. _____

D. _____

7. Name the four major components of a tilting arbor table saw.

_____

_____

_____

_____

8. A miter gauge guides material at angles other than _____ to the blade.

_____

9. Why is a blade guard essential?

_____

_____

10. _____ *True or False?* A rip blade saws across the grain.

11. _____ Use a(n) _____ when the saw cut is shorter than the length of the material.
    A. miter gauge
    B. rip fence
    C. splitter
    D. overhead guide

12. _____ When ripping plywood using a carbide-tipped rip blade, set the blade height approximately _____.
    A. 1″ (25 mm) above the panel thickness
    B. even with the panel thickness
    C. 1/4″–1/2″ (6 mm–13 mm) above the panel thickness
    D. 3/4″ (19 mm) above the panel thickness

13. Sawing with the blade tilted is known as _____.

_____

14. _____ *True or False?* The workpiece dimensions on a beveled edge will be different between the top and bottom faces.

15. _____ creates two or more thin pieces from thicker wood on edge.

_____

16. _____ Grooves cut perpendicular to the grain are known as _____.
    A. ploughs
    B. dados
    C. relief cuts
    D. gullets

17. To cut full-size sheets, most cabinetmakers use a(n) _____ rather than a table saw.

_____

18. _____ The _____ saw is most noted for sawing stock to length.
    A. band
    B. radial arm
    C. scroll
    D. tilting-arbor table

19. When installing a blade and tightening the arbor nut on a saw, why should you avoid overtorquing the nut?

_____

_____

20. The versatility of the radial arm saw comes from its wide range of _____.

_____

21. The radial arm saw motor and blade assembly tilts 45° left and/or right for _____.

_____

22. _____ Choose a _____ saw for cutting large radius curves and large cabinet components.
    A. radial arm
    B. band
    C. scroll
    D. tilting-arbor table

23. _____ Both band saw and scroll saw operations require relief cuts for making curves when _____.
    A. there is a sharp inside or outside curve
    B. the curve changes direction
    C. cabinet parts will be cut from a large piece of stock
    D. All of the above.

24. Identify the parts of the following band saw.

A. _____

B. _____

C. _____

D. _____

E. _____

F. _____

G. _____

H. _____

Delta International Machinery Corp.

25. On a band saw, _____ position and control the blade above and below the table.

_____

26. What is a U-shaped cut?

_____

_____

27. Identify the parts of the following scroll saw.

A. _____

B. _____

C. _____

D. _____

E. _____

Delta International Machinery Corp.

28. Why should you select a scroll saw blade that will have three or more teeth in contact with the wood at all times?

_____

_____

29. Name two types of cuts made with a scroll saw.

_____

_____

30. Why is the scroll saw the only stationary saw capable of easily cutting interior openings?

_____

_____

31. Why should you inspect the cut edges of a workpiece after making a cut?

_____

_____

_____

32. Chip load on a saw blade depends on four factors. Name those factors.

_____

_____

_____

_____

33. List four important specifications for circular blades.

_____

_____

_____

_____

34. Identify the tooth and blade shapes.

A. _____

B. _____

C. _____

D. _____

E. _____

F. _____

G. _____

Carbide-Tipped Blades

Tooth Shape

A   B   C

Blade Shape

D   E

F   G

Goodheart-Willcox Publisher

35. _____ The _____ is where chips accumulate as teeth cut through the material.
    A. dado
    B. hook angle
    C. grind
    D. gullet

36. _____ A _____ saw blade is an endless bonded loop of thin steel with teeth on one edge.
    A. radial arm
    B. scroll
    C. band
    D. circular

37. Identify the band saw blade sets and shapes.

    A. _____

    B. _____

    C. _____

    D. _____

    E. _____

    F. _____

Straight Blades	
Sets	Shapes

A   B   C   D   E   F

Goodheart-Willcox Publisher

38. Identify the scroll saw blades.

    A. _____

    B. _____

    C. _____

    D. _____

A

B

C

D

Goodheart-Willcox Publisher

39. The standard blade length for a scroll saw is _____".

    _____

40. _____ grinding is the required method to sharpen carbide-tipped blades.

    _____

41. _____ When using a table saw, an out-of-square workpiece might indicate a _____ problem.
    A. table
    B. miter gauge
    C. fence
    D. All of the above.

42. Accurate adjustments can be made on a table saw using a(n) _____.

_____

43. _____ Which of the following can be a source of maintenance problems on radial arm saws?
    A. Rust
    B. Lack of lubrication
    C. Excessive torque on levers
    D. All of the above.

44. Name four adjustments that can be made on a band saw.

_____

_____

_____

_____

45. _____ saw blades can be coiled for storage.

_____

46. Check the airflow periodically on _____ saws, because the pump could be damaged or the hose could be loose or broken.

_____

Activity 23A

Using Stationary Power Tools

You are to make a cabinet part that is a finished size of 3/4″ × 16″ × 20″ from a full sheet of 4′ × 8′ plywood. All four corners have a radius of 6″. Number and explain the steps you would take to make this part with stationary power tools.

_____

_____

_____

_____

_____

_____

_____

_____

_____

_____

_____

_____

_____

_____

_____

_____

_____

_____

_____

_____

_____

_____

_____

_____

_____

Name _____

Notes

CHAPTER 24

Surfacing with Hand and Portable Power Tools

Chapter Review

Carefully read Chapter 24 of the text and answer the following questions.

1. _____ *True or False?* You must read the wood grain before planing.

2. _____ Always scrape away excess adhesive from joints _____.
   A. before planing
   B. during planing
   C. after planing
   D. None of the above.

3. Identify the parts of the following bench plane.

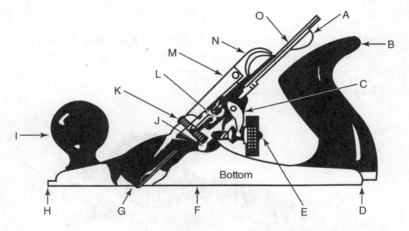

Stanley Tools

A. _____

B. _____

C. _____

D. _____

E. _____

F. _____

G. _____

H. _____

I. _____

J. _____

K. _____

L. _____

M. _____

N. _____

O. _____

4. Name four types of bench planes.

_____

_____

_____

5. _____ *True or False?* The jack plane has less tendency to follow the contours of warped lumber.

6. Bench plane adjustments control the _____, which is the cutter.

_____

7. _____ When planing a surface, you want _____.
   A. thick chips
   B. thin and feathery shavings
   C. sawdust
   D. None of the above.

8. When planing a surface, use _____ strokes with the grain.

_____

9. Which type of plane works well for flattening cupped boards?

_____

10. _____ When planing an edge, _____.
    A. mark a line to which you will cut
    B. plane against the grain figure
    C. hold the plane square
    D. All of the above.

11. Compared to bench planes, a(n) _____ plane is shorter, lighter, and has fewer moving parts.

_____

12. Identify the parts of the following block plane.

greseil/Shutterstock.com

A. _____   D. _____

B. _____   E. _____

C. _____   F. _____

13. _____ True or False? The plane iron on a block plane is set at a lower angle than it is on a bench plane.

14. _____ Scrapers are effective on _____.
    A. edge grain
    B. end grain
    C. face grain
    D. Both A and C.

15. _____ True or False? When scraping, the tool is behind the cutting edge.

Name _____

16. Identify each scraper edge shown.

    A. _____

    B. _____

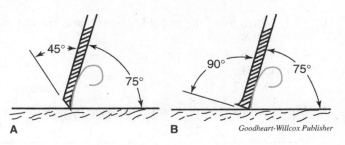

Goodheart-Willcox Publisher

17. List four safety tips to follow when surfacing with hand tools.

_____

_____

_____

_____

18. Identify the parts of the following portable power plane.

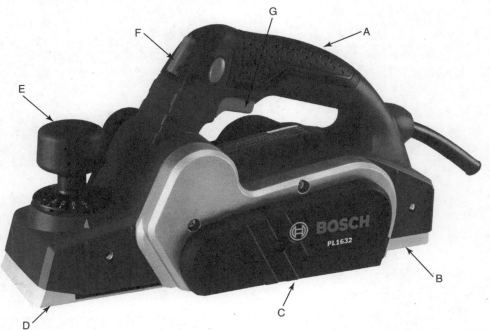

Robert Bosch Tool Corporation

    A. _____   E. _____

    B. _____   F. _____

    C. _____   G. _____

    D. _____

19. What two adjustments are necessary when setting up the portable power plane?

_____

_____

20. _____ When using the portable power plane, the last cut made on a surface should be _____″ or less.
    A. 1/4
    B. 1/8
    C. 1/16
    D. 1/32

21. Why is it important to remove rust from metal surfaces?

_____

_____

 Activity 24A

The Hand Plane

In the space provided, explain why you agree or oppose the following statement: The hand plane has no place in the modern production cabinetmaking shop.

_____

_____

_____

_____

_____

_____

_____

_____

_____

_____

_____

_____

_____

_____

_____

_____

_____

_____

_____

_____

_____

_____

_____

_____

_____

_____

_____

_____

_____

_____

_____

_____

CHAPTER 25

Surfacing with Stationary Machines

Chapter Review

Carefully read Chapter 25 of the text and answer the following questions.

1. Surfacing usually corrects _____ wood.

   _____

2. _____ *True or False?* In wood with cross-grain, the lines formed by the annual rings run parallel to each other the full length of the board.

3. Look at the following diagram. Indicate what type of grain it is and in what direction it is fed during surfacing.

   A. Grain: _____

   B. Direction of feed: _____

   Goodheart-Willcox Publisher

4. _____ The jointer is a multipurpose tool for surfacing _____.
   A. face
   B. edge
   C. end grain
   D. All of the above.

5. _____ When preparing for a jointer surfacing, set the jointer fence at _____°, using a square.
   A. 15
   B. 45
   C. 60
   D. 90

6. When jointing, what is the minimum thickness the material should be? Why?

   _____

   _____

7. _____ *True or False?* When jointing a cupped workpiece, place the cupped side down to prevent the material from rocking.

8. When jointing an edge, check to see that the fence is at a(n) _____° angle to the table.

   _____

   _____

9. Identify the parts of the jointer.

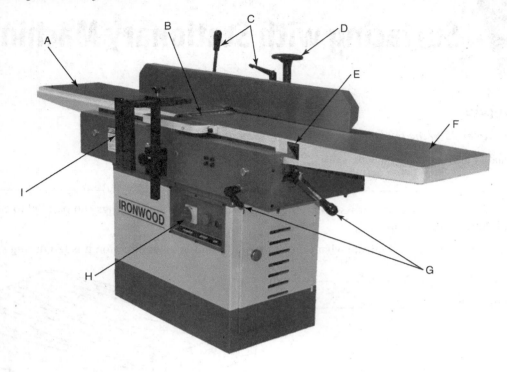

A. _____   F. _____

B. _____   G. _____

C. _____   H. _____

D. _____   I. _____

E. _____

10. When jointing _____, advance the end about 1″ (25 mm) into the cutterhead, then lift and turn the workpiece around.

_____

11. What operation is being shown in the following photo?

_____

Name _____

12. To make the second face of a board parallel to the first, use a(n) _____.

_____

Match each term with the correct description.

13. _____ Reduces friction between the workpiece and the table.

14. _____ Holds stock against the table after the cut is made.

15. _____ Holds the workpiece down and reduces splitting.

16. _____ Additional material removed from the leading or trailing end of a board.

17. _____ Grabs the stock to pull it out from under the cutterhead.

18. _____ Grips the wood to help pull it into the cutterhead.

A. Pressure bar
B. Infeed roller
C. Table rollers
D. Chip breaker
E. Outfeed roller
F. Snipe

19. Identify the parts of the planer.

Stiles Machinery, Inc.

A. _____   E. _____

B. _____   F. _____

C. _____   G. _____

D. _____   H. _____

20. List three steps to prepare a planer for surfacing.

_____

_____

_____

21. _____ *True or False?* For hardwoods, such as oak, the feed rate should be faster than for softwoods, such as pine.

22. _____ When planing glued stock, _____.
   A. use a slower feed rate than normal
   B. use a normal feed rate
   C. use a feed rate faster than normal
   D. None of the above.

23. _____ The minimum thickness for planing stock without a backing board is _____.
   A. 1″ (25 mm)
   B. 3/8″ (10 mm)
   C. 1/4″ (6 mm)
   D. 1/8″ (3 mm)

24. Identify the following machine.

_____

Goodheart-Willcox Publisher

25. _____ Surfacing machines should be kept _____.
   A. clean
   B. properly adjusted
   C. lubricated
   D. All of the above.

26. Before surfacing, help prevent damage to cutting edges by checking for what four items?

_____

_____

_____

_____

27. _____ *True or False?* The sharpness of planer and jointer knives can be determined by listening to the machine.

28. Sparks that fly _____ the wheel indicate a dull edge.

_____

29. Identify the following jointer operation problems.

A. _____

B. _____

C. _____

D. _____

A

B

C

D

Goodheart-Willcox Publisher

30. _____ restores the cutting edge to planer knives.

_____

31. List six adjustments that must be checked after planer knives are sharpened.

_____

_____

_____

_____

_____

_____

_____

■■■■ Activity 25A

Troubleshooting and Correcting Planer Problems

Read the following problems and causes with a planer. Identify at least one solution to each problem.

1. **Problem:** The board will not feed through. **Causes:** Pressure bar too low (most common cause); table rollers too low; insufficient pressure on infeed roller or outfeed roller; cut too deep.

   _____

   _____

   _____

   _____

2. **Problem:** Snipe appears at beginning of board only. **Cause:** Front roller set too high.

   _____

   _____

   _____

   _____

3. **Problem:** Snipe appears at end of board only. **Cause:** Rear table roller set too high.

   _____

   _____

   _____

   _____

4. **Problem:** Clip appears 3–6″ (7.6–15.2 cm) from both ends of the board. **Causes:** Pressure bar set too high; table roller set too high.

   _____

   _____

   _____

   _____

5. **Problem:** Board appears to splinter out. **Causes:** Excessive feed, cutting against grain, chipbreaker too high, or green lumber.

   _____

   _____

   _____

   _____

6. **Problem:** Knives raise grain. **Causes:** Dull knives or green lumber is used.

_____

_____

_____

_____

7. **Problem:** Chip marks appear on stock. **Causes:** Exhaust system not working properly, loose connection in exhaust system, or chips stuck on outfeed roller.

_____

_____

_____

_____

8. **Problem:** Taper across width. **Cause:** Table is not parallel with cutterhead.

_____

_____

_____

_____

9. **Problem:** Glossy or glazed surface appearance on stock. **Causes:** Dull knives or too slow a feed.

_____

_____

_____

10. **Problem:** Washboard surface finish. **Causes:** Knives not set at the same height, too fast a feed rate, or the table gibs are loose.

_____

_____

_____

_____

11. **Problem:** Chatter marks across width of board (small washboard). **Cause:** Table rollers too high (particularly noticeable on thin material).

_____

_____

_____

_____

12. **Problem:** Line on workpiece parallel to feed direction. **Causes:** Nick in knives or a scratch in the pressure bar.

_____

_____

_____

_____

13. **Problem:** Excessive noise. **Causes:** Dull knives, joint on knives too wide, or table roller too high for workpiece thickness.

_____

_____

_____

_____

14. **Problem:** Excessive vibration. **Cause:** Knives not sharpened evenly such that they are different heights.

_____

_____

_____

_____

15. **Problem:** Workpiece twists while feeding. **Causes:** Pressure bar not parallel, table rollers not parallel with table, uneven pressure on infeed or outfeed roller, chipbreaker not parallel, or resin buildup on table.

_____

_____

_____

_____

16. **Problem:** Main drive motor kicks out. **Causes:** Excessive cut, bad motor, or dull knives.

_____

_____

_____

_____

17. **Problem:** Feed motor stalls. **Causes:** Bad motor or lack of lubrication on idlers.

_____

_____

_____

_____

CHAPTER
26 Shaping

Chapter Review

Carefully read Chapter 26 of the text and answer the following questions.

1. _____ Which of the following tools are designed strictly for shaping?
   A. Shapers
   B. Moulders
   C. Routers
   D. All of the above.

2. Identify the parts of the following spindle shaper.

   A. _____

   B. _____

   C. _____

   D. _____

   E. _____

   F. _____

   G. _____

   H. _____

   I. _____

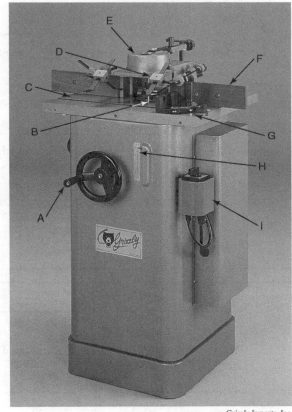

Grizzly Imports, Inc.

3. When hand feeding material into a spindle shaper, it must always be fed against the _____ rotation.

_____

4. Identify each cutter shaper.

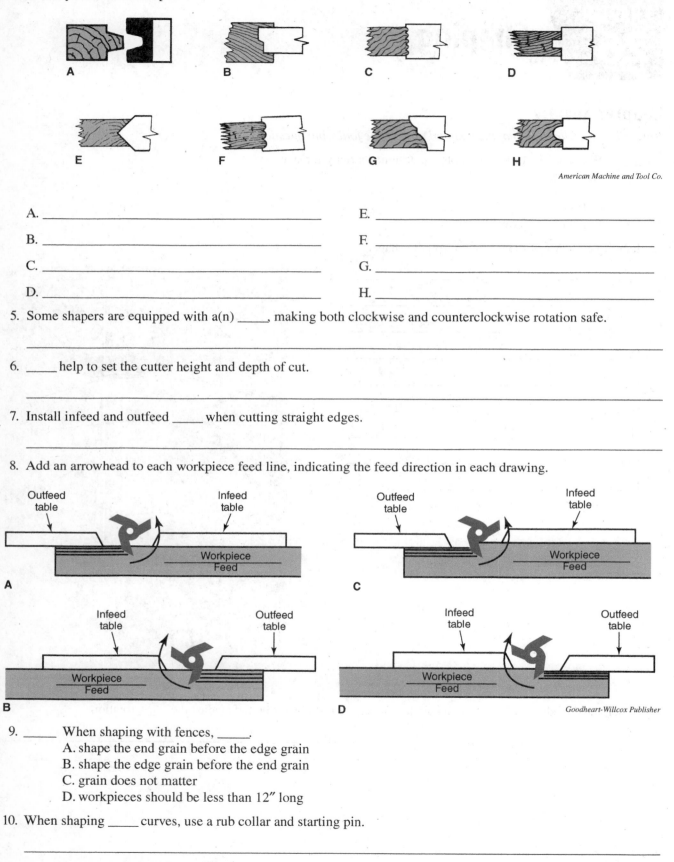

American Machine and Tool Co.

A. _____   E. _____

B. _____   F. _____

C. _____   G. _____

D. _____   H. _____

5. Some shapers are equipped with a(n) _____, making both clockwise and counterclockwise rotation safe.

_____

6. _____ help to set the cutter height and depth of cut.

_____

7. Install infeed and outfeed _____ when cutting straight edges.

_____

8. Add an arrowhead to each workpiece feed line, indicating the feed direction in each drawing.

Goodheart-Willcox Publisher

9. _____ When shaping with fences, _____.
   A. shape the end grain before the edge grain
   B. shape the edge grain before the end grain
   C. grain does not matter
   D. workpieces should be less than 12″ long

10. When shaping _____ curves, use a rub collar and starting pin.

_____

11. _____ are patterns used to duplicate workpieces.

_____

12. _____ The angle jig supports the workpiece at an angle for shaping _____.
   A. bevels
   B. miters
   C. Both A and B.
   D. None of the above.

13. Cove cutting is a shaping operation done on a(n) _____.

_____

14. Identify the parts of the router bit.

   A. _____

   B. _____

   C. _____

Bosch

15. Lowering the router bit into the workpiece is called _____.

_____

16. Identify the typical router bit shapes and the shapes they create.

A   B   C

D   E   F

G   H   I

Bosch

   A. _____   F. _____

   B. _____   G. _____

   C. _____   H. _____

   D. _____   I. _____

   E. _____

17. The _____ router is much like an overarm router, except that the bit is located below the workpiece.

_____

18. Identify the parts of the inverted router.

A. _____

B. _____

C. _____

D. _____

E. _____

F. _____

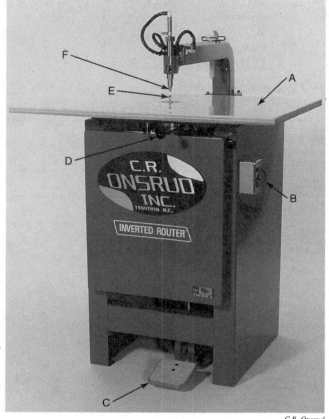

C.R. Onsrud

19. When installing router bits, engage the _____ so the motor shaft will not turn as you tighten the collet.

_____

20. Which part guides the bit in the following diagram?

_____

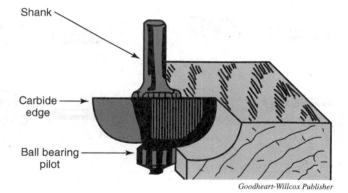

Shank

Carbide edge

Ball bearing pilot

Goodheart-Willcox Publisher

21. To rout a dado, clamp a(n) _____ to the workpiece.

_____

22. _____ Which of the following is used to shape intricate areas?
    A. Laminate trimmer
    B. Motorized rotary tools
    C. Spokeshave
    D. None of the above.

Name _____

23. Identify this tool.

_____

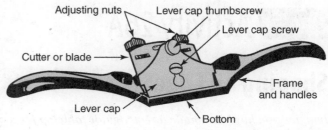

Adjusting nuts Lever cap thumbscrew

Lever cap screw

Cutter or blade

Frame
and handles

Lever cap

Bottom

Stanley Tools

24. Identify this tool.

_____

Patrick A. Molzahn

25. _____ have perforated metal blades with many individual cutting edges.

_____

26. Removing built-up resin on cutters with _____ will prolong the life of the cutting edge.

_____

27. Grind cutters only when the cutting lip is _____.

_____

 **Activity 26A**

Shaping an Edge

Imagine you have just made a rectangular tabletop for a coffee table and you would like to add a profile on the edge. Draw a cross section of one edge of the tabletop, with the edge added. Then explain the procedure you would follow to put the edge on the tabletop. Finally, explain if the procedure would be the same for an oval top.

Cross section

Procedure

_____

_____

_____

_____

_____

_____

_____

_____

Name _____

_____

_____

_____

_____

_____

_____

_____

_____

_____

_____

_____

_____

_____

_____

_____

_____

_____

_____

_____

_____

_____

_____

_____

_____

_____

_____

_____

_____

Notes

CHAPTER 27 / Drilling and Boring

Chapter Review

Carefully read Chapter 27 of the text and answer the following questions.

1. Drilling and boring are _____ processes.

   _____

2. Explain the difference between drilling and boring.

   _____

   _____

   _____

   _____

3. _____ Which of the following drills and bits are the most cost-effective for cabinetmaking?
   A. Carbide-tipped
   B. High-speed steel
   C. Carbon steel
   D. They are all equal.

4. _____ Drilling is done with _____.
   A. hand tools
   B. portable power drills
   C. stationary machines
   D. All of the above.

5. Auger bits are *not* effective for boring holes in _____ grain.

   _____

6. Identify the parts of the following auger bit.

   A. _____

   B. _____

   C. _____

   D. _____

   E. _____

   F. _____

   G. _____

   H. _____

Goodheart-Willcox Publisher

7. _____ *True or False?* Auger bits are fast-speed bits.

8. What type of bit is shown?

   _____

   Irwin

9. Identify the parts of the twist drill.

   A. _____

   B. _____

   C. _____

   Greenlee

10. _____ _____ bits drill a flat-bottomed hole.
    A. Machine spur
    B. Brad point
    C. Spade
    D. Multispur

11. Identify the parts indicated on the following brace.

    A. _____

    B. _____

    C. _____

    D. _____

    E. _____

    F. _____

    G. _____

    H. _____

    I. _____

    J. _____

    K. _____

    Sweep = diameter of swing

    Stanley Tools

12. _____ A _____ is operated with one hand.
    A. brace
    B. hand drill
    C. push drill
    D. drill press

13. The _____ is the most common vertical stationary power drill used by cabinetmakers.

    _____

Name _____

Match each term with the correct description.

14. _____ Effective for drilling pilot holes for screws or nails.

15. _____ Used to drill extra-long holes.

16. _____ Make plugs to cover mechanical fasteners in counterbored holes.

17. _____ Designed for boring holes in wood for pipe and conduit.

18. _____ Have carbide tips to drill holes in concrete and ceramic materials.

19. _____ Angles hole tops to allow flat head screws to sit flush with the surface of the stock.

20. _____ Creates large holes through workpieces.

21. _____ Is flat and has a fairly long brad point center to it.

22. _____ Creates a flatter hole bottom than a machine spur bit creates.

23. _____ Includes combination drills and countersink/counterbore cutters.

24. _____ Twist bit surrounded by a spring-loaded retractable sleeve.

25. _____ Makes holes in concrete for cabinet installation.

26. _____ Has a guide drill and the circumference has a series of saw teeth.

27. _____ Allow user to make long holes with standard drills and bits.

28. _____ Carbide-tipped bits used in boring machines to prepare materials for mounting screws.

29. _____ Spear-shaped, carbide-tipped drill for drilling holes in glass.

A. Circle cutter
B. Star drill
C. Drill extensions
D. Multi-operational bits
E. Brad point bit
F. Spade bit
G. Hole saw
H. Plug cutter
I. Countersink bit
J. Vix bit
K. Multispur bit
L. 32mm system boring bit
M. Drill points
N. Bell hanger's drill
O. Masonry drill
P. Glass drill

30. What type of stationary power machine is designed for the manufacture of European-style cabinetry?

_____

31. Name two types of portable power tools for drilling.

_____

_____

32. What is the purpose of covering vise jaws with wood blocks when drilling and boring?

_____

_____

33. _____ *True or False?* A blind hole goes through the workpiece.

34. Identify the following drill attachments shown.

A. _____

B. _____

A   B

Patrick A. Molzahn

35. Explain how to bore a hole at an angle.

_____

_____

_____

36. Put the following steps for drilling and boring with the drill press in order using 1–9 to indicate the step number.

A. _____ Clamp workpiece to table when drilling large holes, holes at an angle, or when using saw tooth and adjustable bits.

B. _____ Insert the bit in the chuck.

C. _____ Unlock the quill, slide stock to the side of the bit, and adjust the depth stop.

D. _____ Adjust the table.

E. _____ Set the drill press speed.

F. _____ Lower the bit with the feed lever and align the tip with the hole layout marks. Lock the quill.

G. _____ Adjust the guard to within 1/4″ (6 mm) of workpiece.

H. _____ Turn motor on and lower bit into workpiece.

I. _____ Raise bit and turn off machine.

37. _____ Drill holes at an angle by _____.
   A. tilting the machine table
   B. using a fixture
   C. using a jig
   D. Both A and B.

38. Explain two methods for drilling equally spaced holes on the drill press.

_____

_____

_____

_____

_____

_____

39. Before sharpening, check to see that the bit is _____.

_____

40. The preferred method for sharpening an auger bit is with an auger bit _____.

_____

41. _____ *True or False?* Twist drills are difficult to sharpen by hand.

42. _____ Spade bits can be sharpened by _____.
    A. grinding
    B. honing
    C. filing
    D. All of the above.

43. Portable drills need grease in the _____ gears.

_____

Activity 27A

Cabinet Installer

Imagine you are a cabinet installer. Explain in detail what drills and bits you would need in your toolbox and why.

_____

_____

_____

_____

_____

_____

_____

_____

_____

_____

_____

_____

_____

_____

_____

_____

_____

_____

_____

_____

_____

_____

_____

_____

_____

_____

_____

_____

_____

CHAPTER 28

Computer Numerically Controlled (CNC) Machinery

Chapter Review

Carefully read Chapter 28 of the text and answer the following questions.

1. What does *CNC* stand for?

_____

2. _____ is used to move materials and products throughout a facility and for repetitive processes like sanding.

_____

3. CNC machinery is driven by _____.

_____

4. What is shown in the following image?

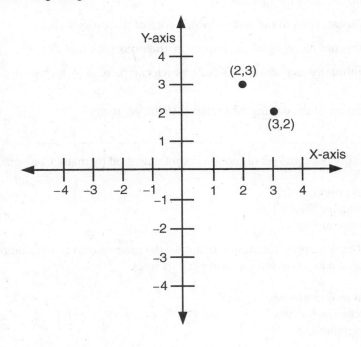

Goodheart-Willcox Publisher

_____

5. Any tool that can be equipped with a(n) _____ for adjusting settings is capable of being controlled by a computer.

_____

6. Name two types of drive mechanisms that are common on CNC machines.

_____

_____

7. _____ *True or False?* Nested base routers require adjustments for varying material sizes.

8. _____ _____ machining centers require that parts be cut to approximate size before machining.
   A. Flat table
   B. Pod-and-rail
   C. Vertical
   D. Cellular

9. The _____ is the brain behind the CNC machine.

   _____

10. List four safety barriers often found on CNC machines that reduce the chance of injury to the user.

   _____

   _____

   _____

   _____

Match each step in the CNC process with the item that performs that step.

11. _____ Visual representation of the product (geometry).

12. _____ Holds the part and executes the operations.

13. _____ Delivers motion instructions to the motors which control the various axes.

14. _____ Converts geometry into meaningful machinery instructions with tool info.

15. _____ Translates machining instructions into G-code which can be read by a specific machine.

A. CAD
B. CAM
C. Post-processor
D. CNC controller
E. CNC machine

16. The tool _____ is the location of the cutting tool relative to the geometry.

   _____

17. _____ _____ is a universally accepted set of motion commands used by many CNC machine tools.
   A. G-code
   B. Conversational programming
   C. Drawing interchange format
   D. Parametric programming

18. _____ _____ is a type of generic programming that allows the programmer to use computer-related features to create programs that can be reused for multiple part sizes.
   A. G-code
   B. Conversational programming
   C. Drawing interchange format
   D. Parametric programming

19. Explain the function of a spoilboard in the CNC machining process.

   _____

   _____

   _____

   _____

20. What is *onion-skinning*?

_____

_____

_____

_____

_____

21. The speed at which the tool moves through the material is known as _____.

_____

22. What is chip load? How does it affect tool life?

_____

_____

_____

_____

23. Calculate the spindle speed in rpm that would need to be programmed for a CNC router to cut a 3/4″ (19.05 mm) thick MDF pattern using a feed rate of 860 IPM (inches per minute) and 1/2″ double-fluted bit in the spindle. Round answer to the nearest 1000. **Note**: use the upper limit chip size from Figure 28-27 in the book.

_____

_____

_____

_____

24. _____ *True or False?* Carbide is the most common material used for CNC tooling.

25. _____ are devices that store tools for quick retrieval during machining.

_____

26. Name two measurements the machine control must know in order to calculate the correct cutting depth and offset.

_____

_____

27. What type of manufacturing is shown in the following image?

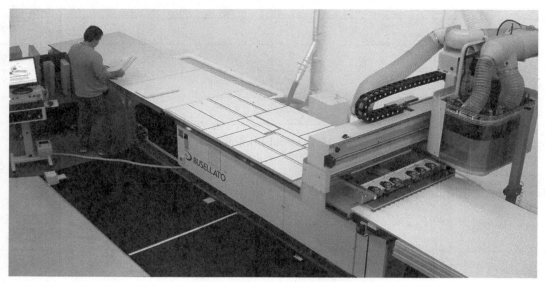

28. The beam saw, shown in the following image, is typical of the machines used in _____ manufacturing.

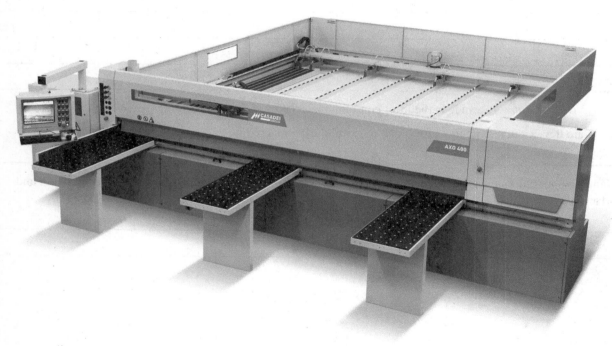

Activity 28A

Figuring CNC Coordinates

List the coordinates of the points that the center point of a 1/2″ diameter CNC router bit would travel to in order to cut the part shown here. Use a straight tool path (refer to Figure 28-28). The Z-axis positioning should not be considered in this activity. Point A is Point 0, 0. Start your cutting at Point A, and proceed clockwise around the piece.

Refer to Figure 28-5 for explanation of the Cartesian coordinate system.

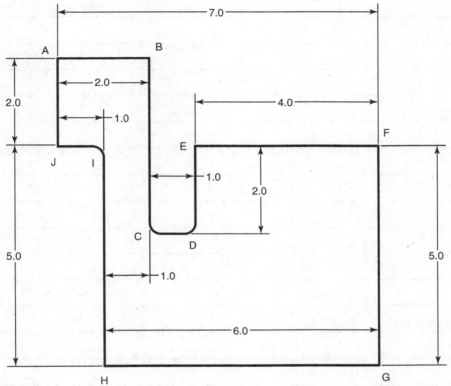

Goodheart-Willcox Publisher

1. Point A: X_____ Y_____

2. Point B: X_____ Y_____

3. Point C: X_____ Y_____

4. Point D: X_____ Y_____

5. Point E: X_____ Y_____

6. Point F: X_____ Y_____

7. Point G: X_____ Y_____

8. Point H: X_____ Y_____

9. Point I: X_____ Y_____

10. Point J: X_____ Y_____

Notes

CHAPTER
29 Abrasives

Chapter Review

Carefully read Chapter 29 of the text and answer the following questions.

1. Abrasives are used by cabinetmakers to _____ surfaces in preparation for assembly or finishing.

   _____

2. _____ abrasives are grains of a natural mineral or synthetic substance bonded to a cloth or paper backing.

   _____

3. _____ abrasives are grains bonded into stones and grinding wheels.

   _____

4. _____ abrasives, such as pumice and rottenstone, are finely crushed abrasives.

   _____

5. Explain how natural abrasives differ from synthetic abrasives.

   _____

   _____

   _____

   _____

   _____

6. Natural abrasives used by cabinetmakers are listed. Describe them.

   Garnet: _____

   _____

   _____

   _____

   _____

   Emery:_____

   _____

   _____

   _____

   _____

Pumice: _____

_____

_____

_____

_____

Rottenstone: _____

_____

_____

_____

_____

Tripoli: _____

_____

_____

_____

_____

7. Some of the synthetic abrasives used by cabinetmakers are listed. Describe them.

Aluminum oxide: _____

_____

_____

_____

_____

Silicon carbide: _____

_____

_____

_____

Industrial diamond:_____

_____

_____

_____

_____

8. _____ size refers to the number of screen holes per linear inch.

_____

9. _____ Which of the following is the heaviest paper backing?
   A. A
   B. C
   C. D
   D. E

10. Which is the finer abrasive and leaves a smoother surface when used: (240) under the CAMI grit no. scale or (P240) under the FEPA grit no. scale?

   _____

11. _____-grade cloth backings are for belt and hand sanding machines.

   _____

12. _____-grade cloth backings are used for flat sanding belts on large production machinery.

   _____

13. Heavy-duty disk and drum abrasives are made with _____ backing.

   _____

14. Abrasive grains are attached to the backing between two separate layers of _____.

   _____

15. _____ On a closed coat abrasive, _____ of the backing is covered with abrasive grains.
   A. the entire surface
   B. 50%
   C. 70%
   D. 80%

16. When bonded by _____, grains of 150 grit or coarser dropped directly on a wet make coating.

   _____

17. Identify each of the following types of flexing.

   A. _____

   B. _____

   C. _____

   D. _____

A

B

C

D

Norton

Match each abrasive form with the correct description.

18. _____ Scored sheet of abrasive bound to a center core with a shaft.

19. _____ Available in diameters of 4 1/2″ through 12″ (114 mm through 305 mm).

20. _____ Coated abrasive bonded to a flexible sponge.

21. _____ Made for arbor-mounted drum and spindle sanders.

22. _____ Range in size from less than 1″ (25 mm) to 52″ (1.32 m) wide.

23. _____ Used for sanding channels, recesses, and bottoms of blind holes.

24. _____ Available in several forms and is used for hand sanding in close quarters.

25. _____ Used for sanding fillets, recesses, and small contours.

26. _____ abrasives are a combination of abrasives and bonding agents.

A. Disk
B. Sleeve
C. Belt
D. Block
E. Flap wheel
F. Pencil
G. Spiral
H. Roll

_____

27. Grinding wheels are most commonly a vitrified bond of _____ oxide grains.

_____

28. Sharpening and honing tools often are made with natural and synthetic _____.

_____

CHAPTER 30 — Using Abrasives and Sanding Machines

Chapter Review

Carefully read Chapter 30 of the text and answer the following questions.

1. Smoothing with _____ is the process of removing surface wood cells to achieve a smooth, blemish-free surface.

   _____

2. _____ Which grain requires the most abrading?
   A. Face grain
   B. Edge grain
   C. End grain
   D. None of the above.

3. Look for _____ caused by planing or scraping before abrading.

   _____

4. How are abrasives selected?

   _____

   _____

   _____

   _____

5. _____ *True or False?* When selecting the proper grit size, each heavier grit size removes the marks of the abrasive before it.

6. Increase grit size by no more than _____ grade numbers at a time below 150.

   _____

7. Sand _____ the grain with the final grit to help hide abrasive scratches.

   _____

8. Explain what is being done in the following photo.

Goodheart-Willcox Publisher

   _____

   _____

9. Name two ways to sand contours.

_____

_____

10. _____ _____ sanders use an abrasive belt that travels around a drive roller and on one or more idler rollers.
    A. Disk
    B. Drum
    C. Dual-action
    D. Edge

11. Identify the parts of the following disk sander.

    A. _____

    B. _____

    C. _____

    D. _____

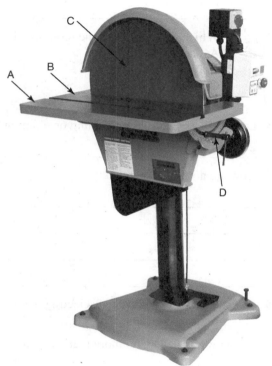

General International Mfg. Ltd.

12. _____ *True or False?* When using the disk sander, only use the half of the disk that rotates upward.

13. What type of sanding machine is shown in the following image?

Courtesy of Tadsen Photography for Madison College

14. _____ Wider _____ sanders are designed to flatten panels.
    A. portable profile
    B. in-line finishing
    C. drum
    D. spindle

15. List five machine- or abrasive-related defects caused by wide belt sanding.

_____

_____

_____

_____

_____

Match each sanding tool with the correct description.

16. _____ Consists of abrasive strips or flat pieces of abrasive mounted on a wheel.

17. _____ Also known as a dual-action sander.

18. _____ Electric motor turns a pair of rollers on which an abrasive belt is mounted.

19. _____ Moves the abrasive back and forth in a straight line.

20. _____ Provides in-line motion to remove machining marks on concave, convex, or flat profiles.

21. _____ Practical where grain direction is not a factor.

22. _____ May be handheld or inserted into portable power drills or stationary drill press.

23. _____ Used with fine abrasives to prepare workpieces for finish or to smooth between finish coats.

A. Portable belt sander
B. Portable finishing sander
C. In-line finishing sander
D. Orbital-action finishing sander
E. Random orbital finishing sander
F. Portable profile sander
G. Flap sander
H. Portable drum sander

24. Name two tools that can be used to remove dust from all moving parts, electrical boxes, and motor vents on sanding equipment.

_____

_____

25. Do not leave _____ disks on sanders after use because they can leave a residue that will cause irregular pressure when sanding.

_____

◤ Activity 30A

Setting Up Shop with Abrasive Machines

Imagine you are setting up a small, two-person cabinetmaking shop. You are responsible for choosing which abrasives and sanding machines you will have in the shop. Choose three machines and the corresponding abrasives you must have. Explain how the machines will be used in your shop.

1. Machine 1: _____

_____

_____

_____

_____

_____

_____

2. Machine 2: _____

_____

_____

_____

_____

_____

_____

3. Machine 3: _____

_____

_____

_____

_____

_____

_____

_____

CHAPTER 31 / Adhesives

Chapter Review

Carefully read Chapter 31 of the text and answer the following questions.

1. _____ Which of the following is a type of adhesive?
   A. Cement or glue
   B. Mastic
   C. Resin
   D. All of the above.

2. _____ occurs by adding a liquid substance that wets the surfaces to be joined and then hardens to withstand stress on the assembly.

   _____

3. Identify each bonding process.

   A. _____

   B. _____

   Porous material

   Porous material

   Layer of
   adhesive enters
   pores

   A

   Layer of
   solvent melts
   plastic

   B   *Goodheart-Willcox Publisher*

4. Select adhesives according to the materials they will _____.

   _____

5. What is shelf life?

   _____

   _____

6. _____ _____ describes the amount of time you have to fit workpieces together after applying adhesive.
   A. Assembly time
   B. Clamp time
   C. Curing time
   D. Shelf life

7. The time after the adhesive sets until the joint reaches full strength is known as _____.

   _____

8. _____ *True or False?* Metal and plastics are porous materials.

9. What is a product data sheet?

_____

_____

10. The _____ surface contact between workpieces, the better the joint.

_____

11. _____ Which of the following affects the durability of a joint?
    A. Temperature
    B. Moisture
    C. Stress
    D. All of the above.

12. Name three forms of wood adhesives.

_____

_____

_____

13. _____ Ease of application is known as _____.
    A. wet tack
    B. sandability
    C. spreadability
    D. gap filling ability

14. Read the following chart and then identify each type of glue being described.

Comparison of Typical Ready-Use Adhesives			
Characteristic	A	B	C
Appearance	Cream	Clear white	Clear amber
Spreadability	Good	Good	Fair
Acidity (pH level*)	4.5–5.0	4.5–5.0	7.0
Speed of Set	Very fast	Fast	Slow
Stress Resistance†	Good	Fair	Good
Moisture Resistance	Fair	Fair	Poor
Heat Resistance	Good	Poor	Excellent
Solvent Resistance‡	Good	Poor	Good
Gap Filling Ability	Fair	Fair	Good
Wet Tack	High	None	High
Working Temperature	45°F–110°F	60°F–90°F	70°F–90°F
Film Clarity	Translucent	Very clear	Clear but amber
Film Flexibility	Moderate	Flexible	Brittle
Sandability	Good	Fair (will soften)	Excellent
Storage (shelf life)	Excellent	Excellent	Good

*pH—glues with a pH of less than 6 are considered acidic and could stain acidic woods such as cedar, walnut, oak, cherry, and mahogany.
† Stress resistance—refers to the tendency of a product to give way under constant pressure.
‡ Solvent resistance—ability of finishing materials such as varnishes, lacquers, and stains to take over a glued joint.

Franklin International

A. _____

B. _____

C. _____

15. Casein and plastic resin are the two basic types of _____ adhesives.

_____

16. What is a thermoset bond?

_____

_____

17. Two-part adhesives use a(n) _____ to harden a resin, forming an adhesive compound.

_____

18. Read the following chart and then identify each type of glue being described.

Comparison of Typical Water-Mixed and Two-Part Adhesives			
Characteristic	A	B	C
Appearance	Cream	Tan	Clear to Amber
Spreadability	Fair	Excellent	Good
Speed of Set	Slow	Slow	†Slow to Fast
Stress Resistance	Good	Good	Excellent
Moisture Resistance	Good	Good	Waterproof
Heat Resistance	Good	Good	Excellent
Solvent Resistance	Good	Good	Excellent
Gap Filling Ability	Fair to Good	Fair	Excellent (nonshrinking)
Wet Tack	Poor	Poor	Poor to Fair
Working Temperature	32°–110°F	70°–100°F	50°F–120°F
Film Clarity	Opaque	Opaque	†Clear to Amber
Film Flexibility	Tough	Brittle	Tough
Sandability	Good	Good	Good
Storage (shelf life)	1 year	1 year	Unlimited (if unmixed)

† Depending on formulation

Franklin International

A. _____

B. _____

C. _____

19. _____ Contact cement works well as an adhesive on all of the following except _____.
   A. wood
   B. the decorative face of plastic laminate
   C. metal
   D. ceramic

Match each type of adhesive with the correct description.

20. _____ A form of thermoplastic adhesive designed to be melted in a hot-glue gun.

21. _____ Used primarily for bonding decorative laminate to particleboard or MDF substrate.

22. _____ Bond unfinished and prefinished plywood, hardboard, and similar panels to wood, metal, and concrete.

23. _____ Also known as super or instant glues.

24. _____ Nontoxic and nonflammable; will not damage lacquered, painted, or varnished surfaces.

25. _____ Nonflammable, dries very fast, develops high strength, and can be applied in a variety of ways.

26. _____ Designed to attach to vinyl trim.

A. Solvent-based contact cement
B. Chlorinated-based contact cement
C. Water-based contact cement
D. Panel adhesive
E. Vinyl-based adhesive
F. Cyanoacrylate adhesive (CA)
G. Hot-melt adhesive

27. Fitting an assembly together before applying glue is known as a(n) _____ run.

_____

28. What is being done in the following photograph?

_____

_____

29. _____ In which of the following situations has glue been applied properly?
A. Small beads of glue ooze from the joint.
B. No glue oozes from the assembly.
C. Glue drips from the joint.
D. None of the above.

30. _____ Glue from an electric gun cools and sets in approximately _____ seconds.
A. 15
B. 30
C. 45
D. 60

31. What process is taking place in the following image?

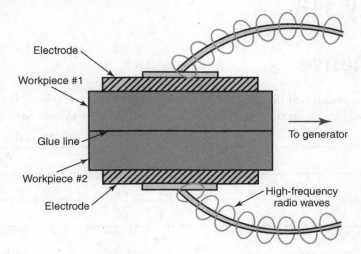

Electrode

Workpiece #1

Glue line

Workpiece #2

Electrode

To generator

High-frequency
radio waves

32. _____ Which of the following adhesives is *not* adaptable for RF gluing?
   A. Urea formaldehyde resin
   B. Cross-linking polyvinyl acetate resin
   C. Hide glues
   D. Aliphatic resin

 Activity 31A

Supply the Adhesive

Imagine you are setting up a small, two-person cabinetmaking shop. You are responsible for ordering essential adhesives for the shop. What three adhesives would you choose and what would you use each adhesive for?

1. Adhesive 1: _____

_____

_____

_____

_____

_____

_____

_____

2. Adhesive 2: _____

_____

_____

_____

_____

_____

_____

_____

3. Adhesive 3: _____

_____

_____

_____

_____

_____

_____

_____

CHAPTER 32 / Gluing and Clamping

Chapter Review

Carefully read Chapter 32 of the text and answer the following questions.

1. List four ways cabinetmakers can use clamps.

   _____

   _____

   _____

2. Identify the following type of clamps.

   _____

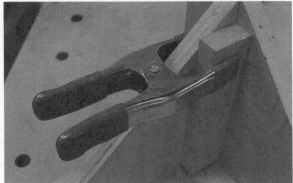

Patrick A. Molzahn

3. Name a disadvantage of using screw clamps.

   _____

   _____

4. Identify the following type of clamp.

   _____

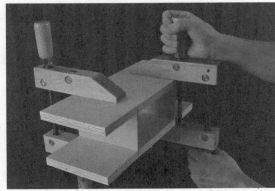

Patrick A. Molzahn

5. Identify the following type of clamp.

_____

Patrick A. Molzahn

6. Identify the following type of clamp.

_____

Bessey Tools North America

7. Identify the following type of clamp.

_____

Patrick A. Molzahn

8. Identify the following type of clamp.

_____

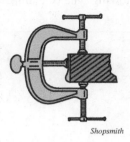

Shopsmith

9. _____ Which of the following clamps are essential for irregular shapes?
   A. C clamps
   B. Web clamps
   C. Band clamps
   D. Both B and C.

10. Identify the following type of clamp.

_____

Bessey Tools North America

11. Identify the following type of clamp.

_____

Patrick A. Molzahn

12. Identify the following type of clamp.

_____

Bessey Tools North America

13. Identify the following type of clamp.

_____

Patrick A. Molzahn

14. Identify the following type of clamp.

_____

Patrick A. Molzahn

15. Identify the following type of clamp.

_____

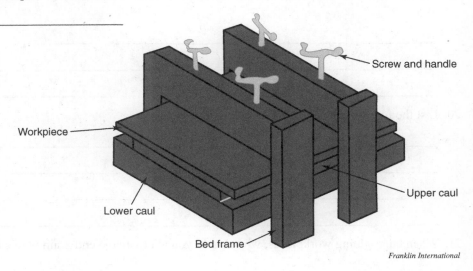

Screw and handle

Workpiece

Upper caul

Lower caul

Bed frame

Franklin International

16. Identify the following type of clamp.

_____

American Tool Companies

17. Identify the following type of clamp.

_____

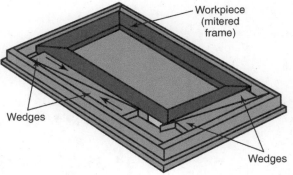

Workpiece (mitered frame)

Wedges

Wedges

Goodheart-Willcox Publisher

18. Backer blocks and _____ protect your product from damage caused by clamp jaws.

_____

19. Explain why you should cut and place wax paper between clamps or backer blocks and the workpiece at a glue joint.

_____

_____

_____

20. List the four primary steps in the clamping procedure.

_____

_____

_____

_____

21. When edge-gluing workpieces, why should you alternate the end grain with each piece?

_____

_____

22. When clamping workpieces face-to-face, the most common problem with hand screw clamps is that the jaws are not aligned _____ to the approximate opening of the assembly.

_____

23. When clamping frames, it is important that you work on a(n) _____ surface.

_____

CHAPTER 33 / Bending and Laminating

Chapter Review

Carefully read Chapter 33 of the text and answer the following questions.

1. _____ When choosing wood for bending, select wood with _____.
   A. straight grain
   B. knots
   C. figured grain
   D. splits

2. _____ *True or False?* Moisture content increases bendability of wood.

3. _____ is specialized plywood available for radius work.

   _____

4. List three items that make wood easier to bend.

   _____

   _____

   _____

5. Wood's attempt to return to its original shape is called _____.

   _____

6. Name two ways to bend lumber while dry.

   _____

   _____

7. Workpieces to be bent dry should pass a(n) _____.

   _____

8. _____ *True or False?* A kerf-bent curve is strong.

9. Identify the following types of curves.

Kerf opens   Kerf closes

A   B

Goodheart-Willcox Publisher

A. _____

B. _____

10. Name the three measurements you must find when kerf bending.

_____

_____

_____

11. _____ bending involves softening, or plasticizing, the lumber.

_____

12. _____ When wet bending, for good bendability, the initial moisture content of wood should be _____.
   A. between 5% and 10%
   B. between 12% and 20%
   C. 100%
   D. None of the above.

13. Identify the following type of wet bending molds.

A

B

Goodheart-Willcox Publisher

A. _____

B. _____

14. _____ When wet bending, pressure should be kept on the bent wood _____.
   A. for one hour
   B. for one day
   C. until it dries
   D. None of the above.

15. Wood _____ is the process of bonding two or more layers of lumber or veneer.

_____

16. List three reasons laminating is done.

_____

_____

_____

_____

_____

17. Give four examples of common laminations.

_____

_____

_____

_____

18. _____ *True or False?* Most curved product laminating is done with successive layers having different grain direction.

19. _____ are used primarily for structural laminations, such as beams and plywood.

_____

20. Look at the following illustration. Which face is best for straight laminations?

Tangential face (face grain)

Radial face (edge grain)

End grain

Goodheart-Willcox Publisher

_____

21. _____ *True or False?* Adhesives for laminating should have a long set time.

22. _____ laminations are done to increase strength and increase thickness.

_____

23. List three ways of making curved laminations.

_____

_____

_____

24. With full surface, one direction laminations, _____ is not reversed, so it is easier to bend.

_____

25. _____ laminations are curves built of rows of solid wood pieces.

_____

 **Activity 33A**

Designing for Bending and Laminating

Design a small product out of wood that makes use of one of the processes discussed in the chapter. The design process should stop at the preliminary idea stage. Dimensions are not necessary. Explain in detail the bending or laminating process you would use in making the project.

Preliminary sketches

Explanation of processes

_____
_____
_____
_____
_____
_____
_____
_____
_____
_____
_____
_____
_____
_____
_____

CHAPTER 34 / **Overlaying and Inlaying Veneer**

Chapter Review

Carefully read Chapter 34 of the text and answer the following questions.

1. Name and describe two types of overlaying.

   _____

   _____

   _____

   _____

2. Name and describe two types of inlaying.

   _____

   _____

   _____

   _____

3. Identify the overlaying or inlaying processes shown in the following illustrations.

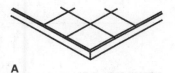      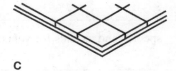   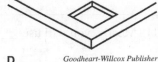

A   **B**   **C**   **D**   *Goodheart-Willcox Publisher*

   A. _____

   B. _____

   C. _____

   D. _____

4. The surface to be decorated is called a(n) _____.

   _____

5. _____ For inlaying, the substrate is normally _____.
   A. plywood
   B. surfaced solid wood workpieces
   C. fiberboard
   D. particleboard

6. Generally, rotary and flat slicing methods produce _____ grain in veneer.

   _____

7. How can you flatten veneer if it warps and distorts?

_____

_____

_____

8. Identify the following tool.

_____

9. _____ When you need time to align veneer, which adhesive is the best choice?
   A. Resin-based, water solvent adhesive
   B. Contact adhesive
   C. Hot melt glue
   D. None of the above.

10. Light, even _____ must be applied to an adhesive coated veneer to create a permanent bond.

_____

11. Mostly, you will use _____ species and exotics for veneering.

_____

12. Look at the following layouts. Identify the matches that can result from them.

A. _____

B. _____

C. _____

13. When cutting veneer, _____ cutting is discouraged.

_____

14. Veneer is placed in a(n) _____ before trimming it with a plane to prevent it from bending.

_____

15. What is being done in the following illustration?

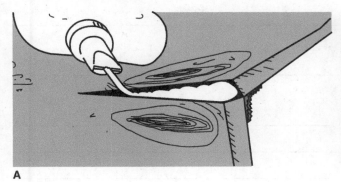

A   B *Goodheart-Willcox Publisher*

A. _____

B. _____

16. Describe two approaches to making repairs to veneers.

_____

_____

_____

_____

_____

_____

_____

17. _____ is an overlaying process in which you cover the edges of manufactured panel products.

_____

18. How does parquetry differ from veneering?

_____

_____

19. List the four steps involved in parqueting.

_____

_____

_____

_____

20. _____ Most often, parquetry involves _____.
        A. circles
        B. curved designs
        C. straight-line geometric shapes
        D. None of the above.

21. When cutting blocks for parquetry, accuracy is important and cuts are made with a(n) _____-tooth blade.

_____

22. Use _____ adhesive when parqueting larger surfaces.

_____

23. Most parqueted surfaces have a protective _____ surrounding the edge.

_____

24. _____ are precut, decorative overlays.

_____

25. List the six steps involved in the inlaying process.

_____

_____

_____

_____

_____

_____

26. Name two ways marquetry pieces can be cut.

_____

_____

27. What is being done in the following illustration?

Depth of cut

Straight bit

_____

28. Describe three ways to prevent inlays from being discolored during finishing.

_____

_____

_____

_____

_____

29. The production sequence for industrial veneering applications used in architectural millwork are similar to what is used in the small shop except the dimensions and _____ is much larger.

_____

_____

30. Describe the operation of a continuous-feed splicer.

_____

_____

_____

_____

Activity 34A

Veneer Chess Board

Explain how you would make a chess board using veneers glued to a substrate material. Describe the materials, including the glue used. Make sure to number your steps.

_____

_____

_____

_____

_____

_____

_____

_____

_____

_____

_____

_____

_____

_____

_____

_____

_____

_____

_____

_____

_____

_____

_____

_____

_____

_____

_____

_____

_____

_____

_____

_____

_____

_____

CHAPTER 35 Installing Plastic Laminates

Chapter Review

Carefully read Chapter 35 of the text and answer the following questions.

1. Name two types of rigid plastic laminates.

   _____

   _____

   _____

2. _____ Which of the following is *not* a good choice for a core material for plastic laminates?
   A. Particleboard
   B. Fiberboard
   C. Solid lumber
   D. Plywood

3. Most cabinet and countertop laminations are done with _____ laminates because of their wear resistance.

   _____

4. Wipe the surface to be covered in laminate with _____ to clean it.

   _____

5. You can score and fracture rigid laminates as you would _____.

   _____

6. When cutting laminate, the saw teeth must enter the laminate on the _____ side on the cutting stroke.

   _____

7. When two pieces of rigid laminate will be butted against each other, an accurate cut can be made with a(n) _____.

   _____

8. _____ is the primary adhesive for laminating.

   _____

9. When applying adhesive to laminates, the temperatures should be no lower than _____°. The relative humidity should be between _____% and _____%.

   _____

10. Generally, you apply a laminate _____ before installing a(n) _____.

    _____

11. _____ removes the sharp corners and extends the wear life of laminate edges.

    _____

12. A warm _____ or a(n) _____ can be used to heat laminate to make it flexible so that it can be laminated around a curve against a core.

    _____

13. What is being done in the following photo?

_____

_____

14. _____ *True or False?* Special contact adhesive must be used when using Kerfkore in laminating products.

 Activity 35A

Laminating an Island Top

Write the procedure that you would use to laminate a kitchen bar counter with plastic laminate HPDL. The countertop has 1 1/2″ edge reinforcement and 4″ radius corners. Describe the materials, including the glue used. Make sure to number your steps.

_____

_____

_____

_____

_____

_____

_____

_____

_____

_____

_____

_____

_____

_____

_____

_____

_____

_____

_____

_____

_____

_____

_____

_____

_____

_____

_____

_____

CHAPTER 36 / Turning

Chapter Review

Carefully read Chapter 36 of the text and answer the following questions.

1. Turning processes produce round parts on a(n) _____.

_____

2. Identify the following two types of turning.

   A. _____

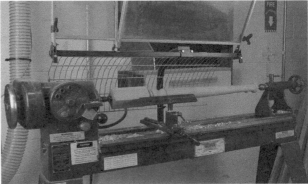

Patrick A. Molzahn

   B. _____

Patrick A. Molzahn

3. Identify the parts indicated on the following standard wood turning lathe.

General International Mfg. Ltd.

A. _____   E. _____

B. _____   F. _____

C. _____   G. _____

D. _____   H. _____

4. What type of lathe is shown in the following photo?

_____

Teknatool U.S.A., Inc.

5. Identify the following turning tools.

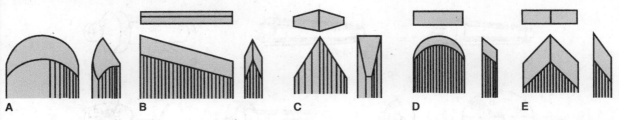

A   B   C   D   E

Patrick A. Molzahn; Goodheart-Willcox Publisher

A. _____

B. _____

C. _____

D. _____

E. _____

6. Look at the following illustrations. Which shows a cutting action and which shows a scraping action?

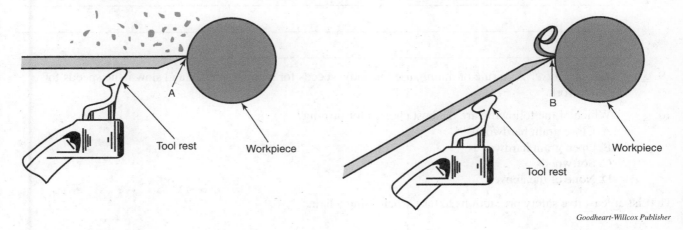

Tool rest Workpiece   Workpiece

A   B

Tool rest

Goodheart-Willcox Publisher

A. _____

B. _____

7. _____ direction affects the smoothness of a surface.

_____

8. Identify the various accessories used for mounting stock.

Delta International Machinery Corp.

A. _____

B. _____

C. _____

D. _____

E. _____

F. _____

G. _____

9. _____ *True or False?* As a rule of thumb, use fast lathe speeds for large diameters and slow lathe speeds for small work.

10. _____ Which of the following are the best choices for turning?
A. Close grain hardwoods
B. Open grain hardwoods
C. Softwoods
D. None of the above.

11. List at least five safety precautions to take when using a lathe.

_____

_____

_____

_____

_____

Match the following terms and descriptions for turning operations.

12. _____ Moving the cutting edge of the turning tool toward the headstock.

13. _____ Feeding the tool toward the centerline of the lathe.

14. _____ Moving the tool away from the centerline of the lathe mostly with the faceplate.

15. _____ Moving the cutting edge of the turning tool toward the tailstock.

A. In
B. Out
C. Left
D. Right

Name _____

16. Describe what is occurring in each of the following illustrations.

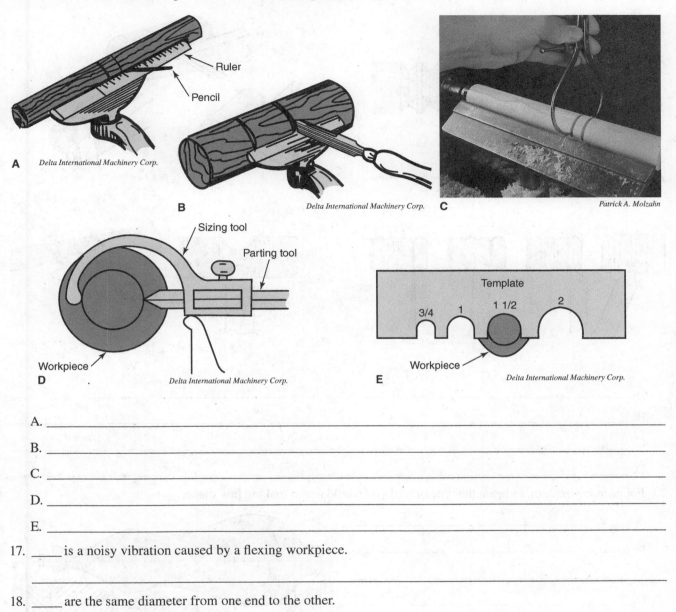

A   *Delta International Machinery Corp.*

B   *Delta International Machinery Corp.*

C   *Patrick A. Molzahn*

D   *Delta International Machinery Corp.*

E   *Delta International Machinery Corp.*

A. _____

B. _____

C. _____

D. _____

E. _____

17. _____ is a noisy vibration caused by a flexing workpiece.

_____

18. _____ are the same diameter from one end to the other.

_____

19. What types of turning are taking place in the following illustrations?

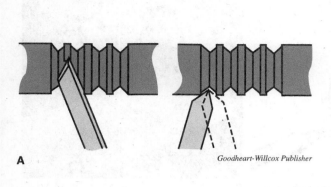

A

Goodheart-Willcox Publisher

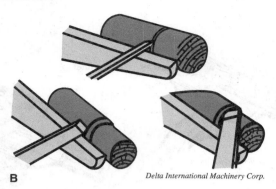

B

Delta International Machinery Corp.

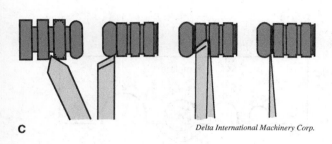

C

Delta International Machinery Corp.

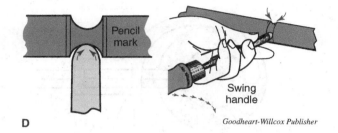

D

Goodheart-Willcox Publisher

A. _____

B. _____

C. _____

D. _____

20. For each key pattern, indicate the type of tool you would use in making this shape.

A. _____

B. _____

C. _____

D. _____

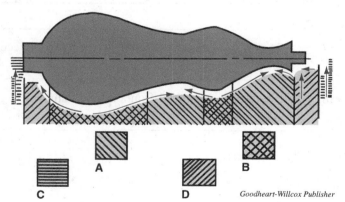

Goodheart-Willcox Publisher

21. _____ Turnings can be made from _____.
    A. a single piece of solid lumber
    B. several layers of glued stock
    C. Both A and B.
    D. None of the above.

22. _____ are glued-up workpieces that are separated after being turned.

_____

23. _____ If you are making spindles in a chair back, which of the following would be the most difficult duplicating practice?
    A. Using a template
    B. Measuring each detail independently
    C. Using a duplicator
    D. Both A and C.

24. You must turn oval spindles _____ times on _____ different centers.

    _____

25. _____ Which of the following are turned while attached to a faceplate?
    A. Bowls
    B. Stool seats
    C. Small round tabletops
    D. All of the above.

26. Material too large for inboard turning is mounted on the _____ side of the lathe.

    _____

27. Dowels and round workpieces are held easily in a lathe _____.

    _____

28. Frequent _____ keeps turning tools sharp and in good condition.

    _____

Activity 36A

Turning Projects

Brainstorm eight turning projects that can be made by turning wood on the lathe. Sketch your ideas in the space provided.

CHAPTER 37 Joinery

Chapter Review

Carefully read Chapter 37 of the text and answer the following questions.

1. One of the most important elements affecting the durability of a product is _____.

_____

2. The direction of grain in a solid wood joint affects the _____ of the joint.

_____

3. _____ Which of the following joints is strongest?
   A. Along the grain joined to along the grain (AG/AG)
   B. Along the grain joined to end grain (AG/EG)
   C. End grain joined to end grain (EG/EG)
   D. They are all of equal strength.

4. _____ Which of the following joints is the strongest?
   A. Radial grain bonded to radial grain
   B. Tangential grain bonded to tangential grain
   C. Radial grain bonded to tangential grain
   D. Both A and B are equal because they shrink at the same rates.

5. Name four joints that are suitable for manufactured panel products.

_____

_____

_____

_____

_____

6. It is best to choose the simplest joint that meets the _____ requirements of the product.

_____

7. List the three main categories of joints.

_____

_____

_____

8. _____ Which of the following is a non-positioned joint?
   A. Mortise and tenon joint
   B. Butt joint
   C. Dovetail joint
   D. None of the above.

9. Explain how positioned joints compare to non-positioned joints.

_____

_____

_____

_____

10. _____ joints have some element in addition to adhesive that helps hold the joint.

_____

11. Identify the following butt joints shown.

Goodheart-Willcox Publisher

A. _____   D. _____

B. _____   E. _____

C. _____

12. _____ *True or False?* Glue blocks make a butt joint weaker.

13. A(n) _____ joint is a slot cut across the grain.

_____

14. A(n) _____ is a slot cut with the grain.

_____

15. Identify the following dado joints.

A. _____

B. _____

C. _____

Goodheart-Willcox Publisher

16. Identify the type of joint used on the following cabinet.

_____

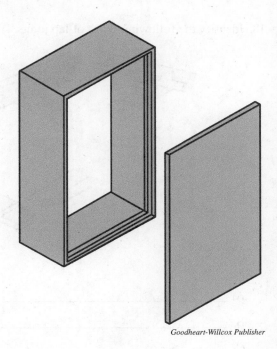

Goodheart-Willcox Publisher

17. Identify the following types of joints.

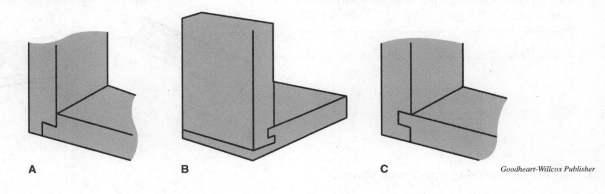

A   **B**   **C**   *Goodheart-Willcox Publisher*

A. _____

B. _____

C. _____

18. Identify the following types of lap joints.

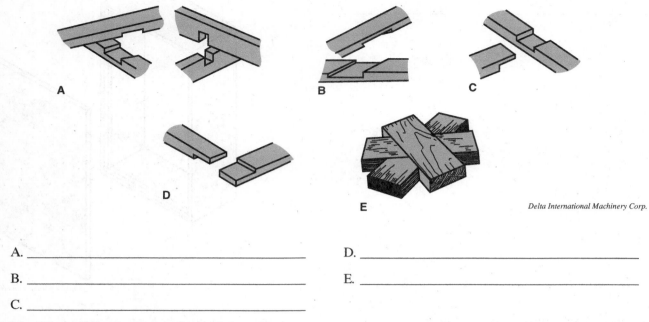

Delta International Machinery Corp.

A. _____   D. _____

B. _____   E. _____

C. _____

19. Identify the following types of miter joints shown.

A. _____

B. _____

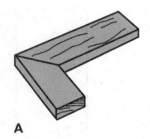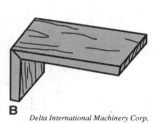

Delta International Machinery Corp.

20. Identify the following type of joint shown.

_____

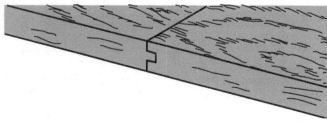

Goodheart-Willcox Publisher

21. Identify the following types of mortise and tenon joints.

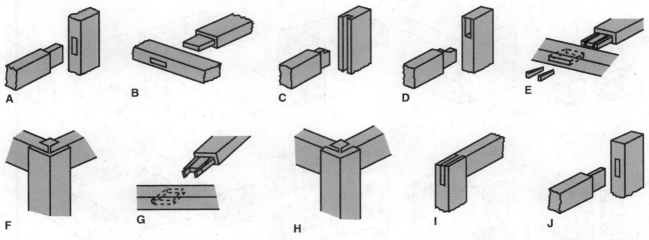

Delta International Machinery Corp.; Stanley Tools

A. _____   F. _____

B. _____   G. _____

C. _____   H. _____

D. _____   I. _____

E. _____   J. _____

22. Identify the following decorative mortise and tenon joints.

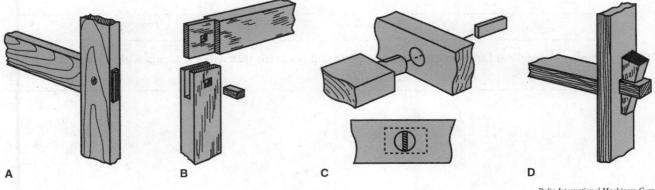

Delta International Machinery Corp.

A. _____   C. _____

B. _____   D. _____

23. Identify the following dovetail joint variations.

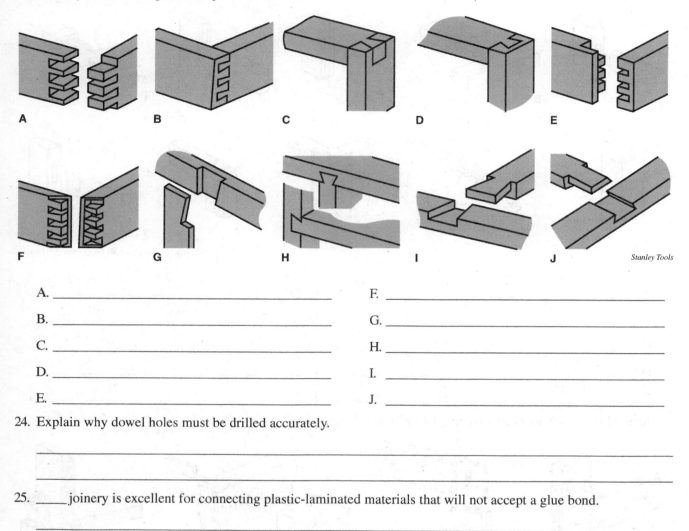

Stanley Tools

A. _____   F. _____

B. _____   G. _____

C. _____   H. _____

D. _____   I. _____

E. _____   J. _____

24. Explain why dowel holes must be drilled accurately.

_____

_____

25. _____ joinery is excellent for connecting plastic-laminated materials that will not accept a glue bond.

_____

26. Identify the various types of spline joints.

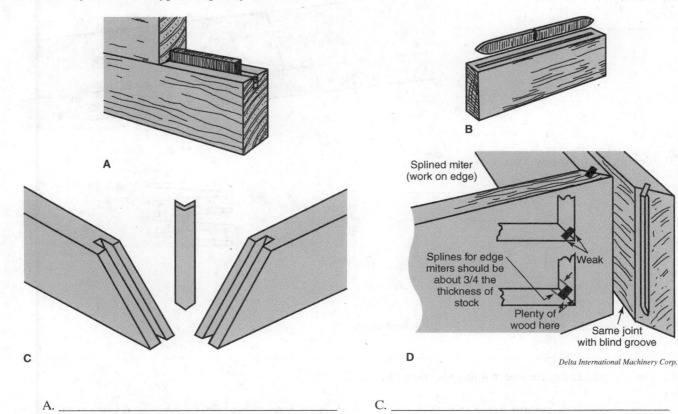

Delta International Machinery Corp.

A. _____   C. _____

B. _____   D. _____

27. Identify the following type of joint.

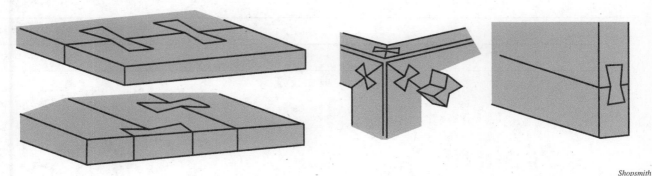

Shopsmith

_____

28. A(n) _____ joint allows the cabinetmaker to use screws to connect end-to-end or edge-to-edge.

_____

29. Identify the following joint.

_____

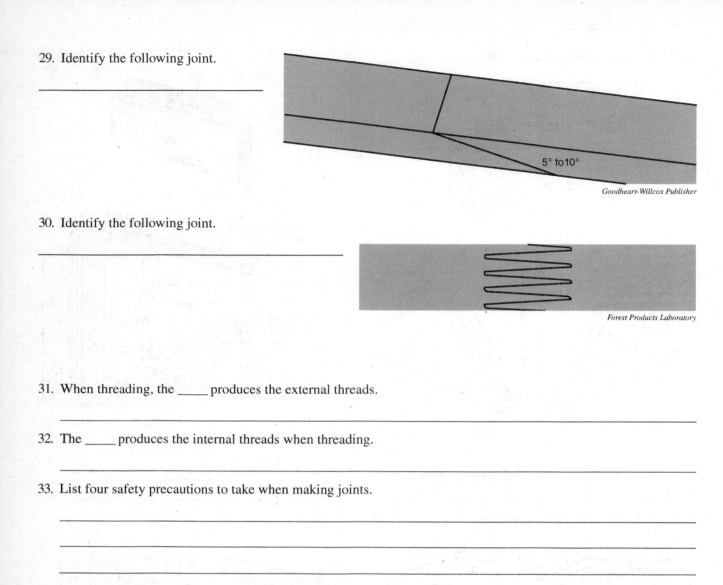

5° to10°

Goodheart-Willcox Publisher

30. Identify the following joint.

_____

Forest Products Laboratory

31. When threading, the _____ produces the external threads.

_____

32. The _____ produces the internal threads when threading.

_____

33. List four safety precautions to take when making joints.

_____

_____

_____

_____

 Activity 37A

Choose a Joint

The following diagram is a plan view of a simple trinket box. What type of joint would you choose to fasten the parts together? Draw the joint in at each corner and explain why you chose that joint.

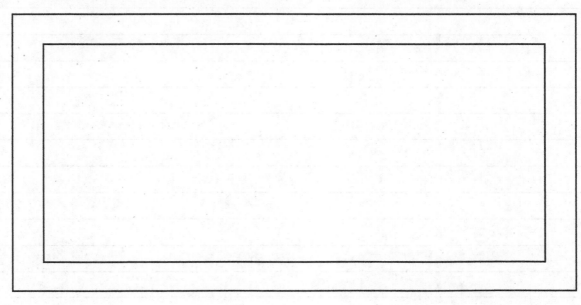

Explanation

_____
_____
_____
_____
_____
_____
_____
_____
_____
_____
_____
_____
_____

CHAPTER 38 — Accessories, Jigs, and Special Machines

Chapter Review

Carefully read Chapter 38 of the text and answer the following questions.

1. Table attachments make saw operations _____.
    A. easier
    B. safer
    C. more accurate
    D. All of the above.

2. _____ help you support large stock.

    _____

3. A(n) _____ supports longer and heavier stock for cross and miter cuts.

    _____

4. _____ keep material against the table and fence and prevent kickback.

    _____

5. Identify the following item and briefly describe its purpose.

Patrick A. Molzahn

    _____
    _____
    _____
    _____

6. Identify the following item and briefly describe its purpose.

Patrick A. Molzahn

_____

_____

_____

7. A(n) _____ consists of a spindle adapter, two guides, and a ring base.

_____

8. _____ hold a workpiece in position and also guide the tool or workpiece.

_____

9. _____ are holding devices that do *not* guide the tool.

_____

10. Identify the following item and briefly describe its purpose.

The Fine Tool Shops

_____

_____

_____

11. Identify the following item and briefly describe its purpose.

Patrick A. Molzahn

_____

_____

_____

12. Jigs for sawing _____ may be a one-piece or an adjustable two-piece device.

_____

13. Identify the following item and briefly describe its purpose.

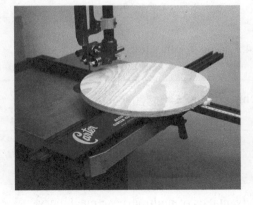

Carter Products Company, Inc.

_____

_____

_____

14. A miter jig can be used instead of a miter gauge for _____° cuts.

_____

15. Identify the following type of fences shown.

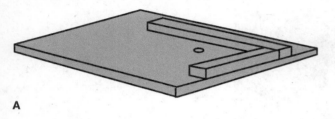

A

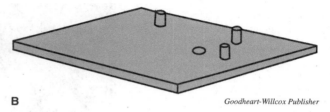

B

Goodheart-Willcox Publisher

A. _____   B. _____

16. Identify the following item.

_____

Adjustable Clamp Co.

17. A(n) _____ trims and fits precise 45° joints for moulding.

_____

18. Explain why a multipurpose machine requires a great deal of planning.

_____

_____

_____

19. List two advantages and two disadvantages of multipurpose machinery.

_____

_____

_____

_____

Activity 38A

Shop Accessories, Jigs, and Special Machines

Name accessories, jigs, and special machines you would like to have in your shop and explain why.

_____

_____

_____

_____

_____

_____

_____

_____

_____

_____

_____

_____

_____

_____

_____

_____

_____

_____

_____

_____

_____

_____

_____

_____

_____

_____

_____

_____

_____

CHAPTER 39 / Sharpening

Chapter Review

Carefully read Chapter 39 of the text and answer the following questions.

1. Which is more dangerous: a dull tool or a sharp tool? Explain why.

   _____

   _____

   _____

2. What are the two main considerations in achieving a sharp edge on a steel tool?

   _____

   _____

3. List three types of tools a cabinetmaker may be required to sharpen.

   _____

   _____

   _____

4. Explain why you should *not* let a tool get too hot in the process of grinding.

   _____

   _____

   _____

5. _____ help hold the tool at a fixed angle to prevent rounding the edge.

   _____

6. _____ Which of the following statements is *not* true?
   A. Grinding is usually the first step in returning a dull edge to the desired angle and removing nicks in a steel blade.
   B. Honing further refines the scratch pattern in the tool edge.
   C. Before proper grinding can be performed, a tool edge must be stropped.
   D. Final stropping of the tool edge is done for final deburring of the sharpened edge.

7. Describe the following abrasive types.

   A. Waterstones: _____

   _____

   _____

   B. Diamond stones: _____

   _____

   _____

C. Oil stones: _____

_____

_____

D. Ceramic stones: _____

_____

_____

E. Abrasive sheets: _____

_____

_____

8. The angle on the cutting face of the tool is referred to as the _____.

_____

9. Explain what to do if your tool becomes too hot to hold while grinding.

_____

_____

10. Explain the steps you would take to sharpen a chisel that has a small nick in the cutting edge. Explain how you would know when it is sharp.

_____

_____

_____

_____

_____

11. List three types of tool blades that are used to hone edge tools by hand.

_____

_____

_____

12. How do most motorized sharpening machines work?

_____

_____

_____

13. Profile knife grinders are used to produce knives for what two tools?

_____

14. _____ *True or False?* Most shops use a sharpening service for their carbide-tipped tooling.

15. _____ *True or False?* Circular saw blades and router bits are usually sharpened in house.

 ## Activity 39A

Shaping an Edge

Draw the edge shape profile of the tools listed below. Diagram angles where applicable. Draw the edge with a hollow ground profile where applicable.

1. Chisel.

2. Cabinet scraper. See **Figure 24-14** in the text for reference.

3. Hand scraper. See **Figure 24-14** in the text for reference.

4. Plane blade.

Notes

CHAPTER 40

Case Construction

Chapter Review

Carefully read Chapter 40 of the text and answer the following questions.

1. List four examples of cases.

   _____

   _____

   _____

   _____

2. With a(n) _____ case, you do not have to edgeband the visible edges of plywood or particleboard.

   _____

3. A(n) _____ case has no face frame; the edges of the top, bottom, and side members are exposed.

   _____

4. _____ The 32mm System is a recent variation of _____ construction.
   A. face frame
   B. frameless
   C. frame-and-panel
   D. web frame

5. Regardless of which method of case construction you choose, the assembled case should be perfectly _____.

   _____

6. List five questions to consider when choosing materials for casework.

   _____

   _____

   _____

   _____

   _____

7. _____ In forming large panels, which of the following is least desirable to use?
   A. Lumber
   B. Plywood
   C. MDF
   D. Particleboard

8. The case _____ refers to the top, bottom, and sides.

   _____

9. Identify the following types of joints shown.

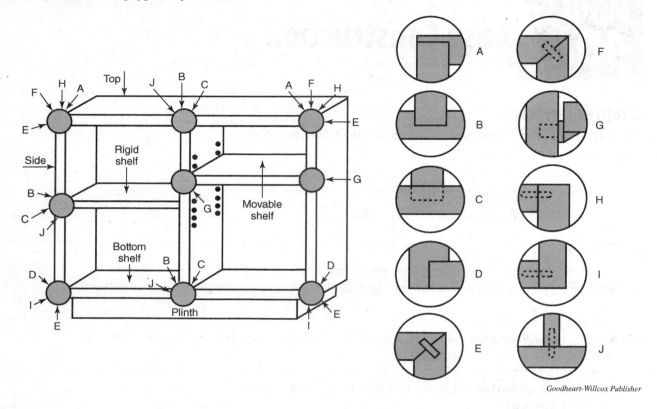

Goodheart-Willcox Publisher

A. _____ F. _____

B. _____ G. _____

C. _____ H. _____

D. _____ I. _____

E. _____ J. _____

10. _____ divide a case into levels for storage.

_____

11. _____ *True or False?* Lumber shelving is less likely to warp than plywood or particleboard.

12. Identify the types of dados used to hold fixed shelves.

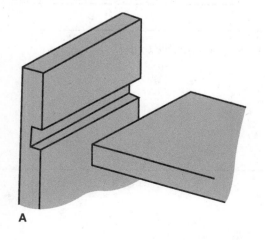

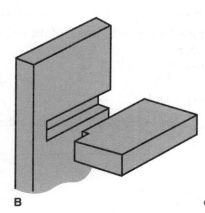

Goodheart-Willcox Publisher

A. _____ B. _____

13. When the edge of a divider is visible on the finished case, cover it with veneer or laminate _____.

_____

14. A(n) _____, or base, provides toe clearance on one or more sides of a case.

_____

15. Tops on cases that open up are generally _____.

_____

16. What is the purpose of the face frame?

_____

_____

17. Horizontal parts of the frame are called _____.

_____

18. Vertical parts of the frame are called _____.

_____

19. Identify types of joints used for face frames.

A   B   C   D   Goodheart-Willcox Publisher

A. _____   C. _____

B. _____   D. _____

20. _____ Successful case assembly requires all of the following *except* _____.
   A. joints that fit precisely
   B. components that are sized correctly
   C. a quick-setting adhesive
   D. components that are sanded smooth

21. Explain why a case should first be assembled without adhesive (a dry run).

_____

_____

_____

22. _____ assembly folds to create a case.

_____

23. _____ The 32mm System _____.
   A. is a standard form of frameless case construction
   B. provides a systematic approach to case assembly and hardware installation
   C. is governed by the minimum spacing of spindles on a multiple-spindle boring machine
   D. All of the above.

24. The _____ distance provides precise alignment of hinge base plates.

_____

25. _____ are 8 mm holes and accept dowels.

_____

26. List three benefits of the 32mm System.

_____

_____

_____

27. With 32mm System case construction, the vertical row of system holes are _____ mm.

_____

28. System holes in the vertical row are spaced _____ mm apart.

_____

29. The vertical row's system holes are spaced _____ mm from the front and rear edge of the panel.

_____

30. Ideally, the height of case side panels should be a multiple of _____ mm, plus the material thickness.

_____

31. _____ Which of the following is *not* typically used to hold a case together?
    A. Dowels
    B. Plates
    C. RTA fasteners
    D. Clamp pads

32. _____ *True or False?* Most 32mm System products look no different from face frame construction that has a full overlay door.

33. Cabinets with visible interiors are often made with veneered or _____ panels.

_____

34. _____ Face frame parts on cabinets can be joined with _____.
    A. dowels
    B. mortise and tenon joints
    C. pocket screw joints
    D. All of the above.

35. For the following kitchen cabinet dimensions, give the metric standard measurements in millimeters.

    A. Base cabinet height with 90 mm-high toe space: _____

    B. Base cabinet depth: _____

    C. Toe space depth: _____

    D. Countertop thickness: _____

    E. Countertop depth: _____

    F. Wall unit depth: _____

    G. Tall cabinet height: _____

    H. Height of pull-out surfaces to allow a knee space: _____

Name _____ Date _____ Class _____

 Activity 40A

Casework Joints

Sketch a drawing similar to the following cabinet provided. Draw the joints you would choose for the application. Then explain why you chose each joint. Remember, you are choosing the application; there is more than correct choice. Make sure you are able to defend your choice.

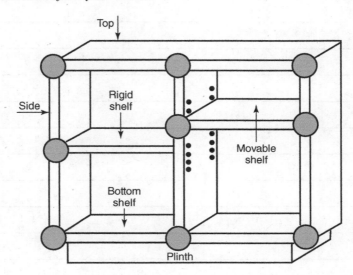

Explanation

_____

_____

_____

_____

_____

_____

_____

_____

_____

CHAPTER 41

Frame and Panel Construction

Chapter Review

Carefully read Chapter 41 of the text and answer the following questions.

1. List two design purposes frame-and-panel assemblies serve.

_____

_____

_____

2. List five engineering purposes frame-and-panel assemblies serve.

_____

_____

_____

_____

_____

_____

_____

3. _____ *True or False?* Frames must be used with panels.

4. Vertical frame side members are called _____.

_____

5. Horizontal frame members are called _____.

_____

6. A vertical piece other than the outside frame is called a(n) _____.

_____

7. Identify the parts indicated on the following frame-and-panel assembly.

A. _____

B. _____

C. _____

D. _____

E. _____

F. _____

G. _____

H. _____

I. _____

J. _____

Shopsmith, Inc.

8. Identify the following frame-and-panel profiles.

Goodheart-Willcox Publisher

A. _____   G. _____

B. _____   H. _____

C. _____   I. _____

D. _____   J. _____

E. _____   K. _____

F. _____

Name _____

9. List four questions you should ask yourself when trying to decide on what type of frame you will make.

_____

_____

_____

_____

10. A stub mortise and tenon is a popular frame joint because the panel groove also serves as the _____.

_____

11. Name two ways stub mortise and tenon frame corners can be reinforced.

_____

_____

12. A haunched mortise and tenon does not require _____.

_____

13. Explain how to make a profiled inside edge.

_____

_____

14. Wood or vinyl stops and plastic retainers hold panels in _____ frames.

_____

15. Raised or beveled and raised panels provide decoration or the look of _____.

_____

16. A(n) _____ panel is made when the surfaces of a product should be flat.

_____

17. A(n) _____ panel lies below the surface of the frame.

_____

18. _____ *True or False?* It is best to apply finish to panels before installing them.

19. What type of frame construction is shown?

_____

20. Identify the items indicated on the following illustrations of joinery in web frame construction.

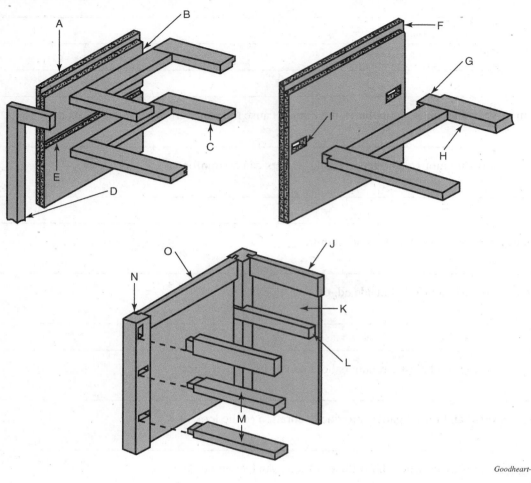

A. _____

B. _____

C. _____

D. _____

E. _____

F. _____

G. _____

H. _____

I. _____

J. _____

K. _____

L. _____

M. _____

N. _____

O. _____

 Activity 41A

Frame-and-Panel Door Design

Imagine that you are designing kitchen cabinets for yourself. You have chosen to use a frame-and-panel type of construction for your doors. Sketch your door design and explain the tooling and procedure you would follow to make the door.

Sketch

Explanation

_____

_____

_____

_____

_____

_____

_____

_____

_____

_____

_____

_____

_____

_____

_____

_____

_____

_____

Cabinet and Furniture Supports

Chapter Review

Carefully read Chapter 42 of the text and answer the following questions.

1. _____ raise cases or furniture above the floor.

   _____

2. What type of bracket feet are shown in the following photo?

   _____

Bracket feet

OZaiachin/Shutterstock.com

3. Each side of a(n) _____ bracket foot is S-shaped.

   _____

4. Name the following leg styles shown.

A

B

C

D

E

F

G

Gerber

A. _____   E. _____

B. _____   F. _____

C. _____   G. _____

D. _____

5. List four basic methods of mounting legs vertically.

_____

_____

_____

_____

_____

6. Use a(n) _____, hanger bolt, and wing nut to secure removable legs to the apron.

_____

7. _____ join a central pedestal with dowels, dovetail joints, and mechanical fasteners.

_____

8. It is easier to join an apron to a(n) _____ section of a leg rather than a tapered section.

_____

9. Identify the following types of decorative legs shown.

A. _____

B. _____

A

B

Goodheart-Willcox Publisher

10. The _____ leg, a distinguishing feature of Queen Anne and French Provincial furniture, often has additions, called ears, extending from each side.

_____

11. Identify the type of legs shown on the following piece of furniture.

_____

Laurel Crown Furniture

12. _____ Which of the following strengthens table and chair supports?
   A. Stretchers
   B. Rungs
   C. Shelves
   D. All of the above.

13. A series of shelves and spindles can be layered and assembled using _____ screws.

_____

14. _____ *True or False?* Posts are longer than legs.

15. Identify the following cabinet support shown.

_____

Dyrlund

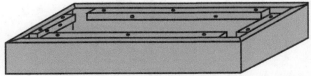

Goodheart-Willcox Publisher

16. Identify the following two items.

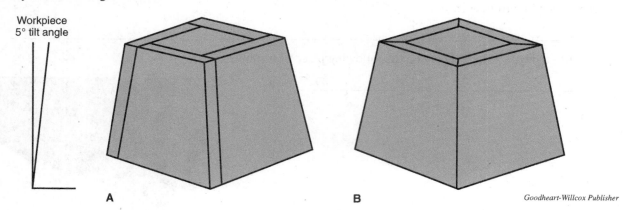

Workpiece
5° tilt angle

A

B

Goodheart-Willcox Publisher

A. _____   B. _____

17. _____ *True or False?* Having the case sides on the floor is the most complex method of support.

18. Identify each of the following items.

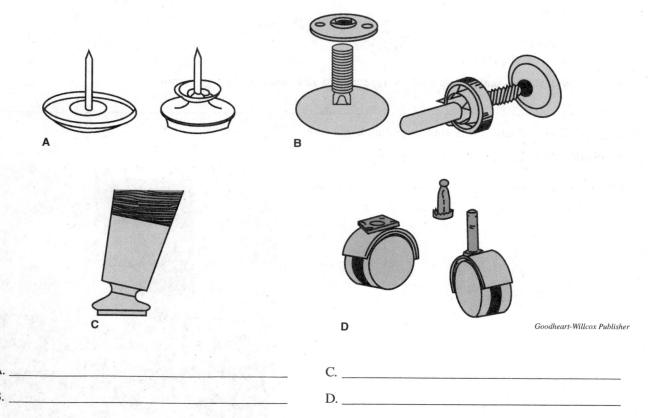

A

B

C

D

Goodheart-Willcox Publisher

A. _____   C. _____

B. _____   D. _____

 **Activity 42A**

Choose Your Support

In this chapter, you have learned about the many ways to support cabinetry and furniture. If you owned a business, you would have to choose the specific techniques and styles to use in the manufacture of your pieces. Draw thumbnail cabinet sketches of supports you would choose to use on an island cabinet, a dining chair, and a dining table to the right of the corresponding cabinetry/furniture. Explain how you reached your decisions.

1. Island cabinet: _____

    _____

    _____

    _____

    _____

    _____

    _____

    _____

2. Dining chair: _____

    _____

    _____

    _____

    _____

    _____

    _____

    _____

3. Dining table: _____

    _____

    _____

    _____

    _____

    _____

    _____

    _____

Notes

CHAPTER 43 / Doors

Chapter Review

Carefully read Chapter 43 of the text and answer the following questions.

1. Name the three basic ways for doors to operate.

   _____

   _____

   _____

2. Large openings, wider than _____″, usually require two hinged doors.

   _____

3. Identify the basic ways that hinged vertical doors mount to a cabinet.

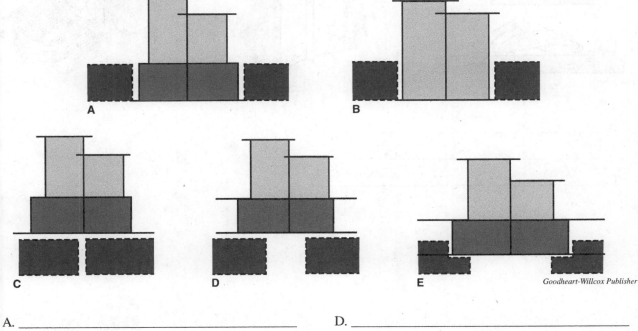

Goodheart-Willcox Publisher

A. _____   D. _____

B. _____   E. _____

C. _____

4. When installing hinges, make sure the pins of the hinges align perfectly to prevent the door from being _____.

   _____

5. Doors over _____″ should have a third hinge to help maintain door alignment.

   _____

6. When selecting fasteners to mount doors, make sure that the screws are _____″ shorter than the door or case thickness.

   _____

7. Identify the treatments on the striker edges on the following double door installations.

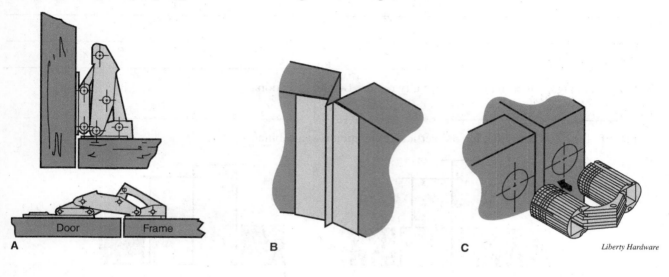

A ___ B ___ C ___ D ___

Goodheart-Willcox Publisher

A. _____   C. _____

B. _____   D. _____

8. Identify the special hinge types used for mounting the following flush doors.

Door   Frame

A ___ B ___ C ___

Liberty Hardware

A. _____

B. _____

C. _____

9. Describe what is being done in the following photos.

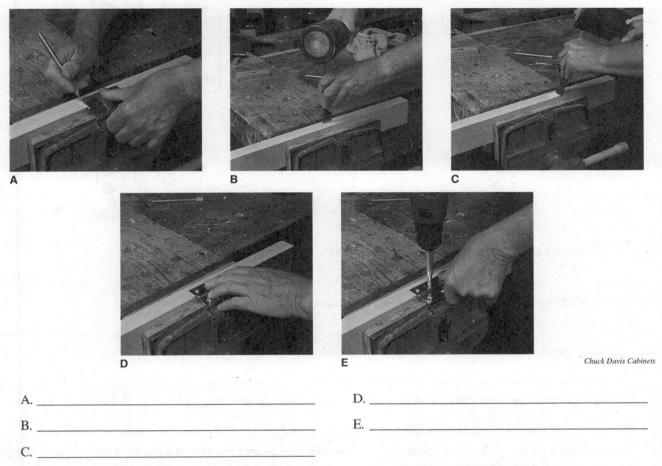

Chuck Davis Cabinets

A. _____   D. _____

B. _____   E. _____

C. _____

10. _____ doors cover some or all of the face frame, or case edges, in frameless cabinetry.

_____

11. Identify the hinge types used to mount overlay doors.

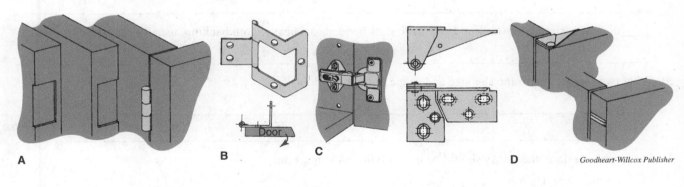

Goodheart-Willcox Publisher

A. _____   C. _____

B. _____   D. _____

12. Identify the following type of hinge.

_____

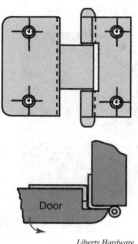

Door

Liberty Hardware

13. _____ hinges are common on lids and hinged-leaf tables.

_____

14. Normally, unframed _____ doors rest in a U-shaped hinge leaf and are held by set screws.

_____

15. _____ doors have two or more panels that glide past each other.

_____

16. Sliding doors are supported and guided by a pair of metal, wood, or plastic _____.

_____

17. The height of a sliding door must measure _____″ less than the distance between the inside of the top track and the top edge of the bottom track.

_____

18. Horizontal sliding _____ must be in place when you fasten the second track.

_____

19. _____ doors consist of narrow wood or plastic slats bonded to a heavy cloth backing, usually canvas.

_____

20. What two factors determine slat size and shape in tambour doors?

_____

_____

21. For a tambour door, the tracks should be nearly twice as long as the _____.

_____

22. _____ Which of the following hold doors closed and can also help open them?
    A. Pulls
    B. Knobs
    C. Catches
    D. Latches

Name _____

23. Describe what is happening in the following photos.

A   B   C   *Patrick Molzahn*

A. _____

B. _____

C. _____

Activity 43A

Door Design Decisions

Contrast which type of hinged door mount configurations (See **Figure 43-2** *in the textbook) is the least expensive to make. Which is the most expensive for the cabinetmaker to make? Explain in detail your choices and why.*

_____

_____

_____

_____

_____

_____

_____

_____

_____

_____

_____

_____

_____

_____

_____

_____

_____

_____

_____

_____

_____

_____

_____

_____

_____

_____

_____

_____

_____

_____

_____

_____

_____

CHAPTER 44 / Drawers

Chapter Review

Carefully read Chapter 44 of the text and answer the following questions.

1. Name two factors that affect drawer construction.

_____

_____

2. _____ *True or False?* Drawers in a chest generally are more shallow near the floor than they are at the top.

3. _____ *True or False?* The drawer type and door mount should be the same.

4. List four ways drawer fronts fit in or against cabinets.

_____

_____

_____

_____

5. What are the five parts of a drawer?

_____

_____

_____

6. _____ Drawer fronts are usually made of _____.
   A. inexpensive, close-grain hardwood lumber
   B. the same material as the cabinet
   C. plywood
   D. hardboard

7. The length of drawer sides varies according to cabinet _____.

_____

8. A(n) _____ drawer bottom rests in a groove cut in the back.

_____

9. A(n) _____ bottom is cut so that the back sits on the bottom.

_____

10. _____ *True or False?* Sand all drawer parts after you cut the joints.

11. _____ To assemble a drawer, apply glue to the _____ and the side-back joints.
    A. front-side joints
    B. drawer front groove
    C. drawer bottom groove
    D. All of the above.

12. What type of joints were used in the following front-side and side-back joints?

_____

DBS drawer box specialties

13. _____ For accuracy, you should make a half-blind dovetail joint using a_____.
   A. dovetail jig
   B. template
   C. router with the appropriate bit
   D. All of the above.

14. What type of dovetail joint has an overhang that butts against the face frame or case edge?

_____

15. _____ *True or False?* Dovetails on lipped drawers are shaped together.

16. What type of joint is shown in the following illustration?

_____

Drawer front

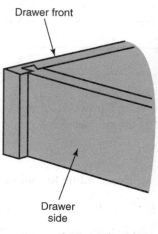

Drawer side

Goodheart-Willcox Publisher

Patrick A. Molzahn

17. What type of joint is shown in the following illustration?

_____

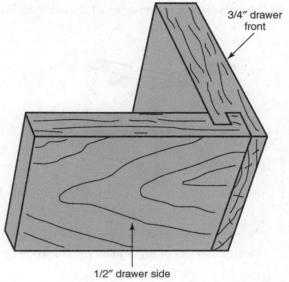

3/4″ drawer front

1/2″ drawer side

Goodheart-Willcox Publisher

18. What type of joints are shown in the following illustrations?

_____

 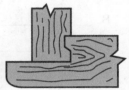

Goodheart-Willcox Publisher

19. Name the typical side-back joint shown in the following illustration.

_____

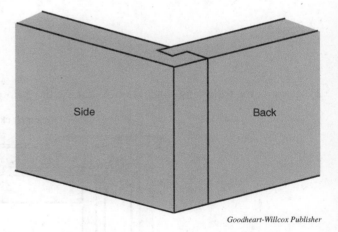

Side

Back

Goodheart-Willcox Publisher

20. Identify the following types of drawer bottoms shown.

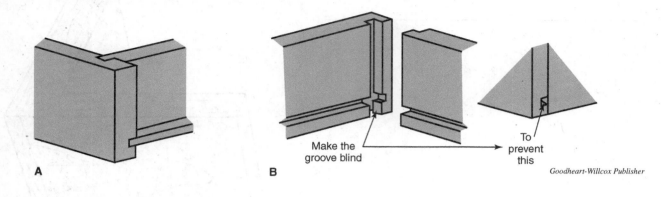

Goodheart-Willcox Publisher

A. _____   B. _____

21. Most producers of high-quality, solid wood cabinetry use _____ construction.

_____

22. Identify the parts indicated on the following construction shown.

A. _____

B. _____

C. _____

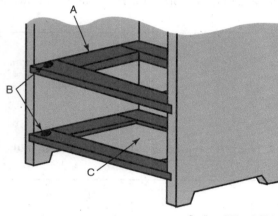

Goodheart-Willcox Publisher

23. Look at the following illustration. What are the purposes of the kicker block and runner?

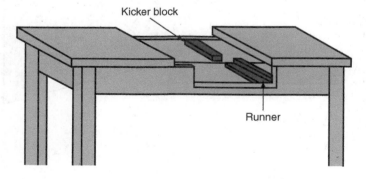

Goodheart-Willcox Publisher

A. Kicker block: _____

B. Runner: _____

24. Identify the items indicated on the following illustration.

    A. _____

    B. _____

    C. _____

    D. _____

25. What are the following items and where are they installed?

_____

_____

26. What type of runner is shown in the following illustration?

_____

27. Why should you be cautious when selecting full-extension slides for free-standing cabinets?

_____

_____

28. List four factors that help determine the number and placement of pulls.

_____

_____

_____

_____

29. What type of drawer system can be used to hide any visible pulls or handles on a drawer?

_____

 Activity 44A

Your Drawer Design

Imagine you are the owner of a cabinetmaking company. You must decide on how your drawers are going to be designed. As you know from the chapter, there are many options. Design exactly how you would choose to build the drawers in your shop in the space provided. It is your choice to draw either a view drawing (top, front, rear, and one side view) or a cabinet drawing. Do not add dimensions; focus on the design configuration and materials. At the bottom of the page, add specifications of design ideas that are difficult to draw, such as specifying what type of fasteners and wood you are using.

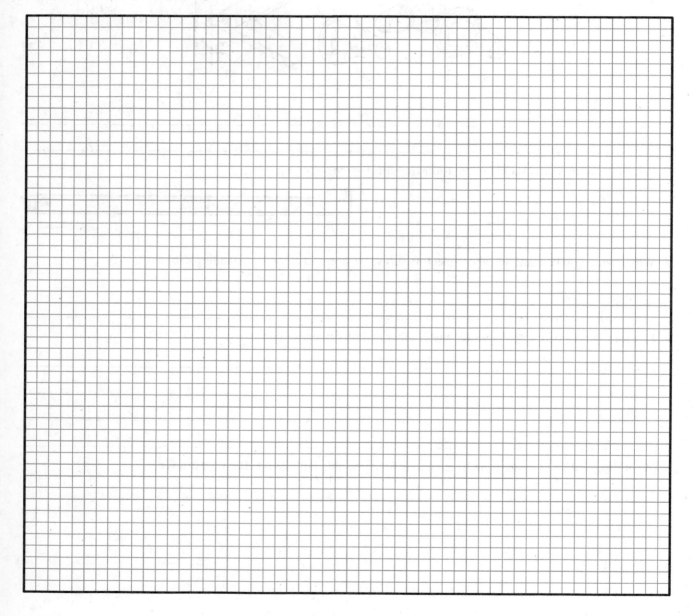

Specifications

_____
_____
_____
_____
_____
_____
_____
_____
_____
_____
_____
_____
_____
_____
_____
_____
_____
_____
_____
_____
_____
_____
_____
_____
_____
_____
_____
_____
_____
_____
_____

Notes

CHAPTER 45
Cabinet Tops and Tabletops

Chapter Review

Carefully read Chapter 45 of the text and answer the following questions.

1. In addition to being attractive, the top of a cabinet or table must be _____ and functional.

   _____

2. _____ *True or False?* Typically, the cabinet or tabletop is the first part to be assembled.

3. _____ *True or False?* Marble and granite are most often used for kitchen and bath countertops.

4. To prevent _____, make the top from a series of wide strips, alternating the faces.

   _____

5. Identify the following three edge treatments.

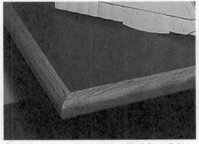

A   *Chuck Davis Cabinets*

B   *Patrick A. Molzahn*

C   *Roth*

   A. _____

   B. _____

   C. _____

6. Identify the type of wood band shown in the following illustration.

   _____

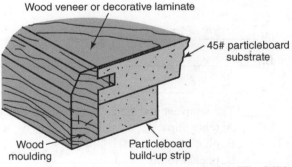

Wood veneer or decorative laminate

45# particleboard substrate

Wood moulding

Particleboard build-up strip

Goodheart-Willcox Publisher

7. What is *edgebanding*?

   _____

   _____

8. _____ Which of the following plastic edgebanding types may be applied with a hand iron?
   A. HPDL
   B. PVC
   C. Melamine
   D. None of the above.

9. While metal was once very popular for kitchen countertops, the material commonly used today to create a durable edge is _____.

_____

10. Identify the following types of hardware used for attaching tops.

A

Desk side   Desk front

B

Glue block (along the grain)

Leg

Apron

Underside of top

Tabletop fastener

C

A. _____

B. _____

C. _____

11. _____ The simplest and least expensive method of attaching a cabinet or tabletop is with _____.
   A. table clips
   B. glue blocks
   C. pocket joints
   D. desktop fasteners

12. _____ tables have one or two sections that hang when the table is not in use.

_____

13. Briefly describe what is happening in each of the following steps of attaching a tabletop with plate joinery.

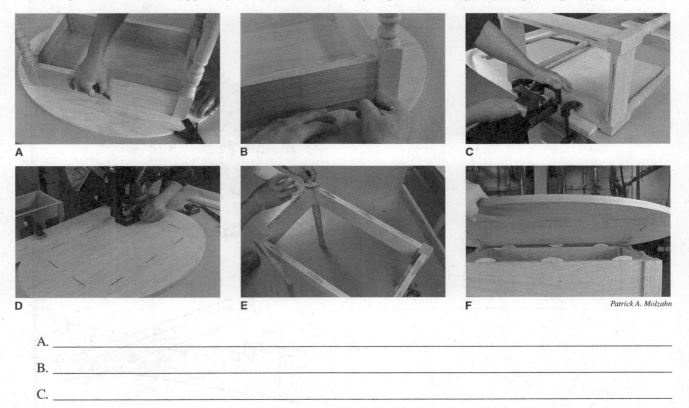

Patrick A. Molzahn

A. _____

B. _____

C. _____

D. _____

E. _____

F. _____

14. What type of joint is shown in the following illustration?

_____

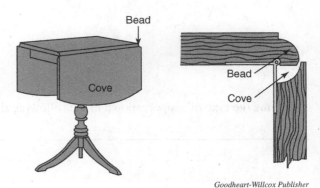

Goodheart-Willcox Publisher

15. Name the type of support shown in the following illustration.

_____

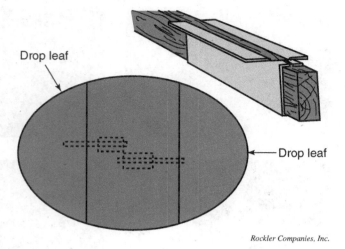

Drop leaf

Drop leaf

Rockler Companies, Inc.

16. Name the type of support shown in the following illustration.

_____

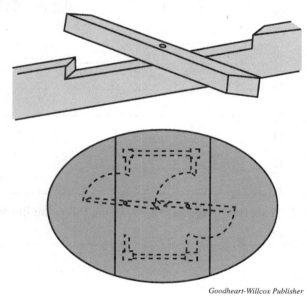

Goodheart-Willcox Publisher

17. Name the type of support shown in the following illustration.

_____

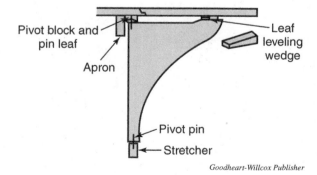

Pivot block and pin leaf

Leaf leveling wedge

Apron

Pivot pin

Stretcher

Goodheart-Willcox Publisher

18. Name the following type of table illustrated.

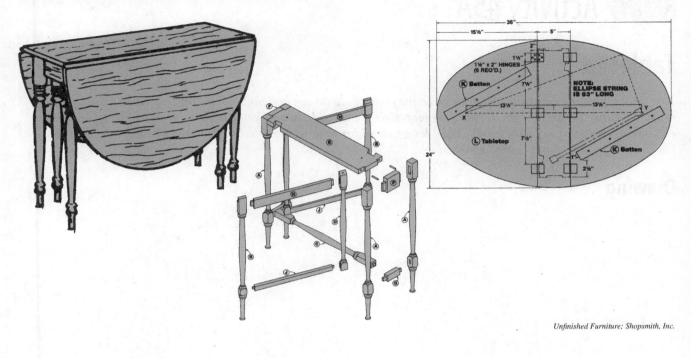

Unfinished Furniture; Shopsmith, Inc.

_____

19. Identify the following item.

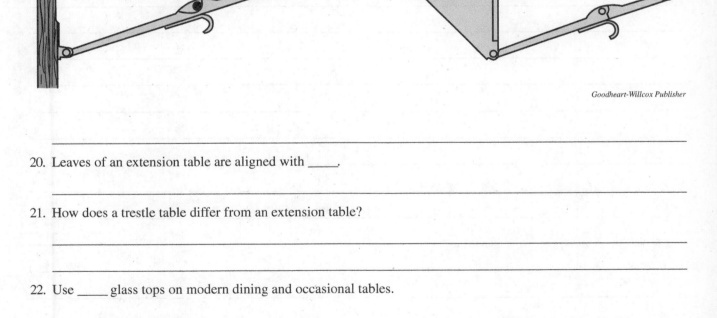

Goodheart-Willcox Publisher

_____

20. Leaves of an extension table are aligned with _____.

_____

21. How does a trestle table differ from an extension table?

_____

_____

22. Use _____ glass tops on modern dining and occasional tables.

_____

23. _____ *True or False?* Glass tops should be frameless.

 **Activity 45A**

Tabletop

Your cabinetmaking company has been given the contract to design and build tables for a restaurant. In the space provided, draw a cross section showing the materials you would use and how the tables would be designed. Then, explain how the tops would be made. Your design should specify what materials are used as substrate, edging, backing, finish, and main covering material.

Drawing

Procedure

_____

_____

_____

_____

_____

_____

_____

_____

_____

_____

_____

_____

_____

_____

_____

_____

_____

CHAPTER 46

Kitchen Cabinet Design and Installation

Chapter Review

Carefully read Chapter 46 of the text and answer the following questions.

1. To meet kitchen needs, the cabinetmaker must consider what four factors?

_____

_____

_____

_____

2. Name eight examples of items that might be stored in kitchen cabinets.

_____

_____

_____

_____

_____

_____

_____

_____

3. _____ refer to countertop or tabletop areas where certain tasks are performed.

_____

4. Name and briefly describe six major work centers in a kitchen.

_____

_____

_____

_____

_____

_____

_____

_____

_____

_____

Copyright Goodheart-Willcox Co., Inc.

May not be reproduced or posted to a publicly accessible website.

273

5. On the following kitchen floor plan, draw in the kitchen triangle.

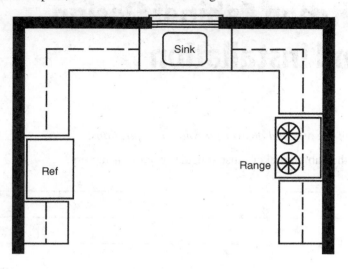

6. What wiring and piping should be in place before cabinets are installed? List three examples for each of the following.

A. Electrical: _____

_____

_____

_____

B. Gas: _____

_____

_____

_____

C. Plumbing: _____

_____

_____

_____

7. An efficient kitchen has a triangle measuring less than _____', but greater than _____'.

_____

8. List five safety factors to consider when planning a kitchen.

_____

_____

_____

_____

_____

_____

9. Identify the following kitchen designs.

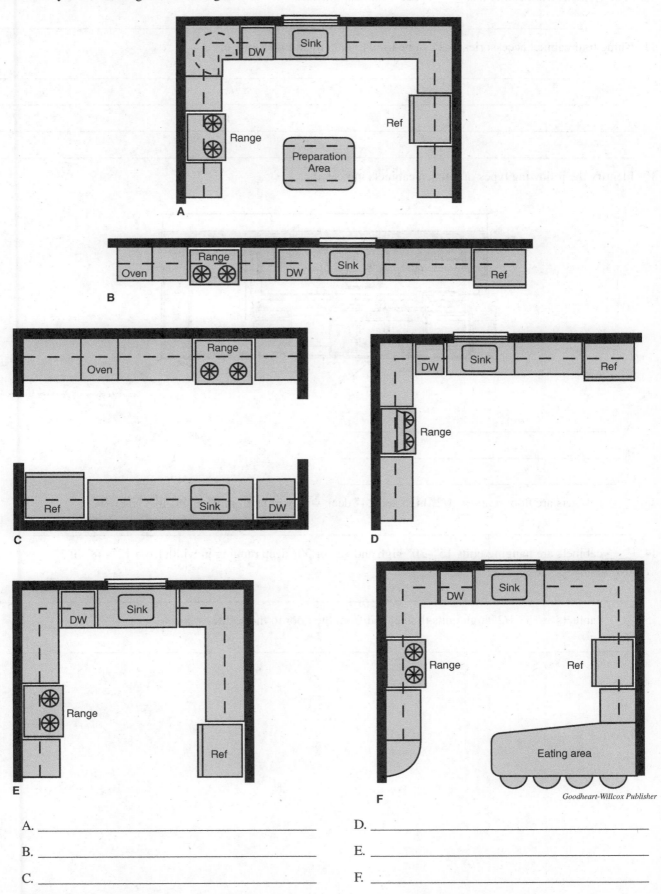

A. _____ D. _____

B. _____ E. _____

C. _____ F. _____

10. A convenient work center has _____ space around appliances.

_____

11. Name four cabinet accessories that can be used to help maximize storage.

_____

_____

_____

_____

12. Identify the following types of kitchen cabinets indicated.

National Kitchen and Bathroom Association

A. _____

B. _____

C. _____

13. _____ cabinets are floor units 34 1/2″ high and 24″ deep with widths from 9″ to 48″ in 3″ increments.

_____

14. _____ cabinets are hanging units 15″–30″ high and 12″ or 30″ deep ranging in width from 12″–48″ in 3″ increments.

_____

15. _____ cabinets are 83 1/2″ high units that extend from the floor to the soffit.

_____

16. Identify the following types of corner cabinets.

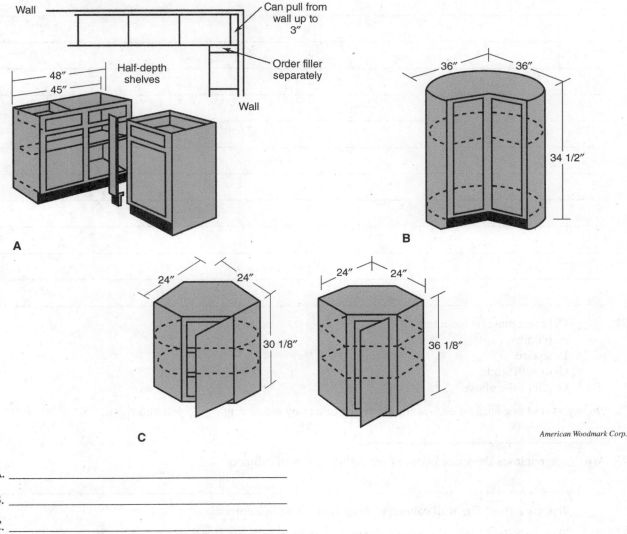

Wall

Can pull from wall up to 3″

Half-depth shelves

48″

45″

Order filler separately

Wall

A

36″ 36″

34 1/2″

B

24″ 24″

30 1/8″

C

24″ 24″

36 1/8″

American Woodmark Corp.

A. _____

B. _____

C. _____

17. _____ cabinets can be found over islands and peninsulas.

_____

18. _____ are open wall units for storing china, crystal, and other display items.

_____

19. A(n) _____ fits between the countertop and wall cabinet.

_____

20. List six tools or types of equipment that you need to install kitchen cabinets and describe how they will be used.

_____

_____

_____

_____

_____

_____

_____

_____

_____

_____

_____

21. _____ Cabinets must be mounted _____.
   A. plumb
   B. square
   C. to wall studs
   D. All of the above.

22. Once you find one stud on each wall, locate the others by measuring _____″ left and right.

_____

23. A(n) _____ encloses the space between the ceiling and wall cabinets.

_____

24. _____ *True or False?* The wall cabinet to hang first is one in a corner.

25. _____ *True or False?* Lay out cabinet heights from the lowest point on the floor.

26. _____ *True or False?* Begin installing base cabinets in the center.

27. Place _____ behind and under cabinets to account for irregular walls and floors.

_____

28. The _____ is installed once the base cabinets are fully secure.

_____

29. When cutting the countertop to length, leave _____″ to _____″ of overhang.

_____

Name _____ Date _____ Class _____

 **Activity 46A**

Kitchen Cabinet Checklist

Pretend you are designing a kitchen for your "dream home." Complete the following checklist to help you determine your kitchen requirements.

Checklist

Before you visit your kitchen specialist, fill out this handy checklist designed to help you determine your own personal kitchen requirements.

Cabinetry

Cabinet Style _____   Wood Type (or laminate) _____
Color _____

Appliances

Electric or Gas Range _____   Built-in Oven _____
Electric or Gas Cooktop _____   Built-in Double Oven _____
Downdraft Cooktop _____   Microwave Oven _____
Grill _____   Built-in Oven/Microwave Combination _____
Electric or Gas Burner/Grill Combination _____   Ventilation System _____
Refrigerator _____   Trash Compactor _____
Freezer _____   Waste Disposer _____
Dishwasher _____   Other: _____

Special Options

- Butler's Pantry
- Snack Bar
- Appliance Garage
- Planning Desk
- Bake Center

- Center Work Island
- Range Hood
- Decorative Glass Door Inserts
- Special Sink

- Second Sink
- Custom Beam
- Custom Appliance Panels
- Plate Rail

- Other: _____

Accessories

(See our Accessories Guide for the complete selection of kitchen accessories.)

Wall Cabinets

- Microwave Cabinet
- Open Shelf Cabinet
- Wall Wine Rack

- Spice Rack
- Corner Lazy Susan
- Tambour Cabinet

- Corner Tambour Cabinet
- Wall Quarter Circle Cabinet

- Pigeonhole
- Hinged Glass Doors
- Other: _____

Base Cabinets

- Cutlery Divider
- Tray Divider
- Bread Box
- Pull-Out Chopping Block

- Pull-Out Table
- Tote Trays
- Wastebasket
- Pull-Up Mixer Shelf
- Base Hamper

- Tilt-Out Soap Tray
- Corner Lazy Susan
- Adjustable Roll-Out Shelves

- Visible Storage Rack Baskets
- Other: _____

Tall Cabinets

- Tray Divider
- Base Hamper

- Visible Storage Rack/ Baskets
- Built-in Oven Cabinet

- Multi-Storage Cabinet
- Adjustable Roll-Out Shelves

- Other: _____

Other Design Considerations

Sink and Fixtures _____   Type of Wall Coverings _____
Special Window Treatment _____   Space Arrangement Changes _____
Type of Flooring _____   _____
Type of Countertop _____   Special Design Needs _____
Type of Lighting _____   _____

 Activity 46B

Measuring Appliances

Visit an appliance store or the appliance department of a department store. For each appliance, record the brand/ make, model number, width (left to right), depth (front to back), and height. You may want to bring along a tape measure.

Freestanding Range

Brand/Make: _____

Model Number: _____

Width (Left to right): _____

Depth (Front to back): _____

Height: _____

Drop-In Range

Brand/Make: _____

Model Number: _____

Width (Left to right): _____

Depth (Front to back): _____

Height: _____

Countertop Cooktop

Brand/Make: _____

Model Number: _____

Width (Left to right): _____

Depth (Front to back): _____

Height: _____

Built-In Wall Oven

Brand/Make: _____

Model Number: _____

Width (Left to right): _____

Depth (Front to back): _____

Height: _____

Microwave

Brand/Make: _____

Model Number: _____

Width (Left to right): _____

Depth (Front to back): _____

Height: _____

Built-In Dishwasher

Brand/Make: _____

Model Number: _____

Width (Left to right): _____

Depth (Front to back): _____

Height: _____

Refrigerator

Brand/Make: _____

Model Number: _____

Width (Left to right): _____

Depth (Front to back): _____

Height: _____

Sink

Brand/Make: _____

Model Number: _____

Width (Left to right): _____

Depth (Front to back): _____

Height: _____

Other

Brand/Make: _____

Model Number: _____

Width (Left to right): _____

Depth (Front to back): _____

Height: _____

Other

Brand/Make: _____

Model Number: _____

Width (Left to right): _____

Depth (Front to back): _____

Height: _____

CHAPTER 47

Built-In Cabinetry, Wall Paneling, and Mantels

Chapter Review

Carefully read Chapter 47 of the text and answer the following questions.

1. Storage units, often custom-made to fit special space restrictions, are called _____.

   _____

2. Give six examples of typical needs for added storage in a home.

   _____

   _____

   _____

   _____

   _____

   _____

3. Give three examples of where added workspace might be needed in a home.

   _____

   _____

   _____

4. List five general rules that you should follow when designing built-in storage.

   _____

   _____

   _____

   _____

   _____

5. _____ *True or False?* Custom-made bathroom vanities can follow the guidelines and standard sizes given for kitchen cabinets, but are higher.

6. _____ Which of the following can be utilized for storage space?
   A. Attics
   B. Stairwells
   C. Wall units
   D. All of the above.

7. What are three built-in activity areas that require utilities?

   _____

   _____

   _____

8. List seven utility and storage factors that must be considered when designing a room to support laundry activities.

_____

_____

_____

_____

_____

_____

_____

_____

9. Plan for _____ to prevent moisture buildup.

_____

10. _____ is an alternative to plastered or gypsum board-covered walls.

_____

11. Identify the following three types of millwork paneling.

A   **B**   **C**   *California Redwood Association*

A. _____

B. _____

C. _____

12. _____ *True or False?* When installing paneling, mouldings should be added over joints.

13. Horizontal millwork paneling attaches directly to _____.

_____

14. Place horizontal _____ strips if you will be installing vertical millwork paneling.

_____

15. List the steps necessary to estimate the number of sheets of paneling that will be needed for a project.

_____

_____

_____

_____

_____

_____

_____

16. When installing vertical paneling, leave about _____″ at the floor.

_____

17. _____ *True or False?* Begin installing horizontal paneling at the ceiling with the tongue edge up.

18. Begin installing _____ panels in a corner.

_____

19. _____ *True or False?* Millwork paneling is generally much easier to install than sheet paneling.

20. _____ *True or False?* Install backing vertically for vertical sheet paneling.

21. After you press paneling with adhesive to the wall, what are two reasons to pull it away for a short time?

_____

_____

22. Identify the steps used to install paneling.

A

B

C

D

E

F

Weyerhauser

A. _____

B. _____

C. _____

D. _____

E. _____

F. _____

23. Identify the following panel installations.

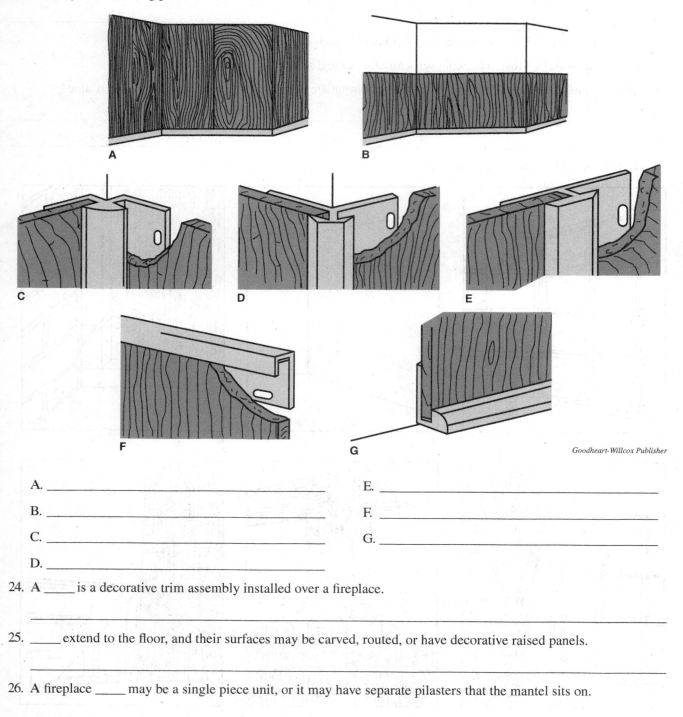

A. _____   E. _____

B. _____   F. _____

C. _____   G. _____

D. _____

24. A _____ is a decorative trim assembly installed over a fireplace.

_____

25. _____ extend to the floor, and their surfaces may be carved, routed, or have decorative raised panels.

_____

26. A fireplace _____ may be a single piece unit, or it may have separate pilasters that the mantel sits on.

_____

 **Activity 47A**

Add Built-in Cabinets to Your Home

Find a good place in your home to add built-in cabinets or shelves. Draw a plan view of the room (looking down from the ceiling) and an elevation view (eye level looking straight at the wall where you would add the built-in cabinets or shelves).

Plan View

Elevation View with Addition

Notes

CHAPTER 48 / Installing Moulding and Trim

Chapter Review

Carefully read Chapter 48 of the text and answer the following questions.

1. _____ Installing mouldings and trim is a task normally reserved for the _____.
   A. cabinetmaker
   B. framing carpenter
   C. finish carpenter
   D. drywall installer

2. _____ mouldings are primarily vertically oriented.
   _____

3. _____ mouldings refer to mouldings that run horizontally.
   _____

4. _____ mouldings are installed horizontally where the floor meets the wall.
   _____

5. Mouldings that are used to surround and strengthen doors and windows are called _____ mouldings.
   _____

6. Mouldings applied where walls and ceilings meet are known as _____ mouldings.
   _____

7. _____ and _____ are horizontal mouldings that run along the walls.
   _____

8. Two main styles of moulding shapes are _____ and _____.
   _____

9. _____ *True or False?* Stock mouldings cannot be combined with existing mouldings.

10. What must be done when a certain, specialized moulding shape is not available?
    _____
    _____
    _____
    _____

11. _____ *True or False?* There is no difference in the material of stain grade mouldings and paint grade.

12. Describe two advantages of using plastic moulding.
    _____
    _____
    _____

13. Because the trim carpenter must carry and use tools from room to room, the most important tool the carpenter must have is a good quality tool belt. List the tools and materials that should be in that tool belt.

_____

_____

_____

_____

_____

_____

_____

_____

14. The _____ has become the most important power tool for the finish carpenter when cutting moulding.

_____

15. _____ *True or False?* While most finish carpenters still carry a hammer, the majority of trim is applied with pneumatic nail guns.

16. _____ *True or False?* Keeping a jobsite clean is only important when a client is around.

17. _____ The preferred way to clear dust is by _____.
   A. sweeping
   B. blowing it out with an air compressor
   C. blowing it out with a leaf blower
   D. vacuuming

18. _____ A power tool or extension cord should *never* be used when _____.
   A. the ground plug is missing
   B. the outer insulation has been nicked
   C. Both A and B.
   D. Neither A nor B.

19. _____ Which of the following statements is *true*?
   A. The trim carpenter has no business inspecting the rough frame because the trim carpenter did not perform the rough frame work.
   B. The rough frame should be inspected to verify opening sizes and double check for plumb, level, and square.
   C. It is a waste of time to inspect the rough frame because door openings and walls are always plumb and square.
   D. Nowadays, the same person who builds the frame also does the trim work.

20. _____ Proper installation of base moulding calls for the inside corner to be _____.
   A. cut with a coping saw
   B. miter cut with a miter saw
   C. left slightly loose for adjustments later
   D. Both B and C.

21. Moulding can be attached with what three types of fasteners?

_____

22. _____ *True or False?* Construction adhesive may be used to attach mouldings to framing.

23. _____ *True or False?* Window and door casings are good examples of standing trim.

24. _____ *True or False?* Preassembled moulding is better suited for paint-grade trim.

25. The two methods of casing a door are _____ and _____.

_____

26. Explain the *reveal* as it relates to installing casing to the jamb of a door or window.

_____

_____

_____

_____

27. Explain how to make slight adjustments on a miter saw cut angle without changing the miter saw setting.

_____

_____

_____

_____

28. The two methods of casing a window are _____ and _____.

_____

29. _____ *True or False?* Curved mouldings cannot be cut on a power miter saw.

30. _____ moulding is the most difficult to cut of the running mouldings.

_____

31. _____ *True or False?* Running mouldings differ from standing mouldings in that they are cut slightly long of the measurement so that pressure is maintained on the joints.

32. What are two ways to cut crown moulding?

_____

33. The following is a list of the major mouldings. State whether each is a running or a standing moulding and then describe the function of the moulding.

A. Base moulding: _____

_____

B. Crown moulding: _____

_____

C. Door casing: _____

_____

D. Window casing: _____

_____

E. Chair rail moulding: _____

_____

34. Which of the following are good work habits? List all that apply.

A. Using a helper.
B. Making cuts at one time when you can.
C. Calling out measurements to a second helper to cut.
D. Working with a miter saw on high scaffold work.
E. Using a digital measuring device.
F. Using a second miter saw to lessen angle changes on one saw.
G. Establishing a clear communication system.
H. Planning your work.

_____

35. List the three levels of quality for finish carpentry as they relate to visual aesthetics put forth by The Architectural Woodwork Institute (AWI) in AWI 0620.

_____

 Activity 48A

Running Base

The following diagram shows the plan view of two rooms. Assume you were to run the base moulding and the door casings. The dashed lines represent the running base moulding. Decide what joints are needed and draw them in the plan view with labels. Your drawing should demonstrate which pieces go through to the wall and which pieces butt up to trim. A schedule of joints is provided to guide your drawings. Your trim pieces come in 16' lengths so you will need a scarf joint at least every 16'. Estimate the distance for the scarf joint.

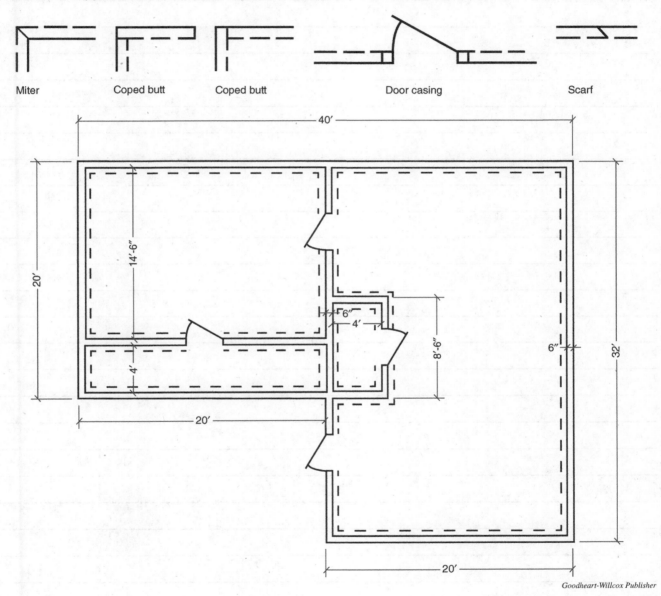

Miter   Coped butt   Coped butt   Door casing   Scarf

Notes

CHAPTER 49

Furniture Project Ideas

Chapter Review

Carefully read Chapter 49 of the text and answer the following questions.

1. _____ Always process duplicate parts of desks _____.
   A. as needed
   B. at the same time
   C. one at a time
   D. None of the above.

2. _____ When making a clock, choose the desired movement _____.
   A. after the clock is designed
   B. just before finishing the outside
   C. during the design stage
   D. None of the above.

3. _____ Few chairs are stable if the legs are _____.
   A. perpendicular to the seat
   B. at an angle to the seat
   C. at different angles between the front and back
   D. None of the above.

4. List three parts of a typical bed.

   _____

   _____

   _____

5. _____ support the box springs and mattress.

   _____

6. What is the typical size of a full-type, cotton or foam mattress?

   _____

7. Identify the parts indicated on the following illustration.

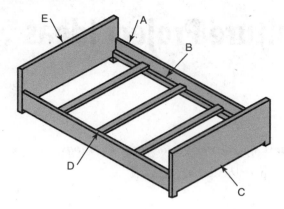

Goodheart-Willcox Publisher

A. _____   D. _____

B. _____   E. _____

C. _____

8. A(n) _____-side water bed mattress is simply a vinyl inner tube that must be placed within a solid frame.

_____

9. Name two ways water beds can be supported.

_____

_____

10. The _____ protects the wall, provides decoration, and sometimes offers storage.

_____

11. _____ prevent bedspreads from slipping off the bed and often provide decoration.

_____

12. A(n) _____ mirror is supported in a stand.

_____

13. _____ are used to separate work areas, reduce drafts, or provide privacy.

_____

14. Look at the following illustration. It is an example of _____ furniture.

_____

Unfinished Furniture

Match the following descriptions with the corresponding bed type.

15. _____ Has corner posts much higher than the headboard.

16. _____ Often includes curtains on the sides for privacy with a wood cap top at each corner.

17. _____ The mattress does not sit on legs.

18. _____ Has a visible straight wooden frame.

19. _____ Stacked one over the another, usually twin size.

20. _____ Nested beds, usually twin size, where the lower bed rolls out on casters from storage under the other bed.

A. Bunk

B. Canopy, arched

C. Canopy, tester

D. Platform

E. Poster

F. Trundle

 Activity 49A

Selecting Furniture

In the space provided, mount pictures from newspaper or magazine furniture ads of furniture you would like to have in your dream bedroom. (Make sure to get approval before cutting up a magazine or journal.)

CHAPTER 50 / Finishing Decisions

Chapter Review

Carefully read Chapter 50 of the text and answer the following questions.

1. List four safety precautions to take when applying finishing materials.

   _____

   _____

   _____

   _____

   _____

   _____

   _____

   _____

2. Finishing involves applying the proper _____ materials to assembled products.

   _____

3. _____ A finish protects wood, wood products, and metal from _____.
   A. moisture
   B. harmful substances
   C. wear
   D. All of the above.

4. According to the chart in **Figure 50-2** in the textbook, what type of finish should be applied to American elm wood?

   _____

5. List the five steps involved in finishing.

   _____

   _____

   _____

   _____

   _____

6. Unwanted natural coloring (mineral deposits, blue stain, etc.) can be removed by _____.

   _____

7. _____ Which of the following involves spattering the surface of wood with a dark colorant before applying the final topcoat?
   A. Scorching
   B. Distressing
   C. Imitation distressing
   D. None of the above.

8. List five methods for applying coating materials.

_____

_____

_____

_____

_____

9. List the four steps in applying topcoating.

_____

_____

_____

_____

10. _____ changes the color of the wood.

_____

11. _____ contain dyes and resins that are almost totally absorbed into the wood.

_____

12. _____ contain insoluble powdered colors that bond to the wood surface with resins.

_____

13. A(n) _____ is a thinned coat of sealer that often is applied before pigment stains.

_____

14. _____ is a liquid or paste material that is used to fill pores.

_____

15. A(n) _____ is a thin, clear coat that fills the pores of the wood and serves as a base coat.

_____

16. _____ prepares surfaces for opaque coatings.

_____

17. _____ Which of the following is *not* a penetrating finish?
    A. Linseed oil
    B. Phenolic resin oil-based coatings
    C. Shellac
    D. Wax

18. Name one advantage and one disadvantage of using penetrating finishes over built-up finishes.

_____

_____

_____

_____

19. _____ finishes do not penetrate the wood; they form a film on the surface.

_____

20. Describe the difference between gloss built-up finish and satin built-up finish.

_____

_____

_____

_____

21. Only use semigloss and satin finishes as the _____ coat.

_____

22. Name two stages of the drying time of a topcoating.

_____

_____

23. _____ are ink transfers bonded to the surface.

_____

24. _____ involves buffing the surface with compounds and, generally, is the last step in the finishing process.

_____

25. _____ Which of the following provides a visual effect of texturing while the surface remains smooth?
    A. Mottling
    B. Antiquing
    C. Gliding
    D. Marbelizing

26. Usually, clear lacquer is applied to hardware and other metal products to prevent _____.

_____

27. _____ Which of the following finishes can be dissolved with alcohol?
    A. Shellac
    B. Lacquer
    C. Enamel
    D. Varnish

28. In the space provided, list the steps you would take to produce a stained, clear satin finish on a cabinet made of an open grain hardwood.

_____

_____

_____

_____

_____

_____

_____

_____

_____

_____

 **Activity 50A**

Decisions! Decisions! Decisions!

Making the right decisions when it comes to what finishing method to use on a particular wood piece will come with experience. The information found in this chapter can speed up that process. Using that information, decide how you would finish the pieces listed here. Use the charts provided in the text and refer back to chapters that describe the particular type of wood used to make the piece. Defend your decisions and contrast them with some of the methods that you did not use. In other words, not only tell why you chose a particular method, but also tell why you did not choose another.

1. Small, aromatic cedar chest: _____

_____

_____

_____

_____

_____

2. Cabinet made of maple veneer plywood and solid-wood face frame: _____

_____

_____

_____

_____

_____

3. Solid walnut coffee table: _____

_____

_____

_____

_____

_____

_____

CHAPTER 51 / Preparing Surfaces for Finish

Chapter Review

Carefully read Chapter 51 of the text and answer the following questions.

1. The method used to prepare a surface depends on the final _____.

   _____

2. List at least four safety precautions to take when preparing surfaces to apply a finish.

   _____

   _____

   _____

   _____

   _____

3. _____ *True or False?* Remove surface defects after applying a finish.

4. _____ Dents can often be raised with _____.
   A. abrasives
   B. a drop of water
   C. stain
   D. None of the above.

5. Name four products that can repair chips, scratches, and voids.

   _____

   _____

   _____

   _____

6. _____ consists of real wood flour in a resin and works well for small defects.

   _____

7. _____ are cellulose fiber fillers.

   _____

8. _____ *True or False?* Glue and wood dust mixed together make an effective patch for large cracks and holes.

9. Precolored _____, made of wax, fills small holes or cracks.

   _____

10. _____ is melted to fill defects in wood.

   _____

11. Identify the following tool.

_____

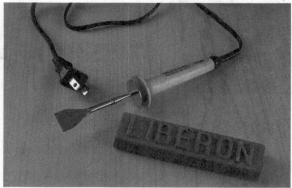

12. Why do some cabinetmakers recommend filling defects for opaque finishes after the first coat?

_____

_____

13. _____ Excess dry glue should be removed with _____.
    A. a hand scraper or cabinet scraper, followed by abrasive paper
    B. water
    C. your fingernail
    D. a chisel

14. Oil spots can be dissolved and removed with _____ or lacquer thinner.

_____

15. _____ is helpful in removing mildew, blue stain, and clamping stains.

_____

16. _____ is the process of swelling wood fibers and any last dents, dimples, or other pressure marks.

_____

17. _____ *True or False?* Higher-density MDF is the easiest to work with when finishing.

18. What is occurring in the following photo?

_____

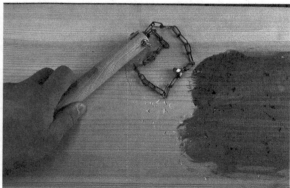

Name _____

19. What is occurring in the following photo?

_____

Patrick A. Molzahn

20. What is occurring in the following photo?

_____

Patrick A. Molzahn

Activity 51A

What Tools to Use?

You are in the business of making fine custom cabinets for homes. Make a list of the supplies and tools you must have on hand to prepare the surfaces for the finish you apply. Explain how you would use each item.

_____

_____

_____

_____

_____

_____

_____

_____

_____

_____

_____

_____

_____

_____

_____

_____

_____

_____

_____

_____

_____

_____

_____

_____

_____

_____

_____

_____

_____

_____

CHAPTER 52 / Finishing Tools and Equipment

Chapter Review

Carefully read Chapter 52 of the text and answer the following questions.

1. Name five finish application practices.

_____

_____

_____

_____

_____

2. List three guidelines to consider when selecting a finish application method.

_____

_____

_____

3. _____ Applying finishes with brushes _____.
   A. involves a limited amount of equipment compared to spraying
   B. is fast compared to spraying
   C. Both A and B.
   D. None of the above.

4. List four features to consider when selecting brushes.

_____

_____

_____

_____

5. _____ *True or False?* Brushes with synthetic bristles are suitable for applying shellac and lacquer.

6. _____ Brushes with natural bristles are suitable for applying _____.
   A. oil-based finishes
   B. water-based finishes
   C. Both A and B.
   D. None of the above.

7. Identify the parts of the paintbrush.

A. _____

B. _____

C. _____

D. _____

E. _____

F. _____

G. _____

Stanley Tools

8. To dip a brush in finish is called _____.

_____

9. _____ When dipping a brush into finish, cover the bristles _____.
   A. more than 60% of the length
   B. no more than 50% of the length
   C. the entire length
   D. None of the above.

10. _____ refers to applying an even film of finish.

_____

11. Describe the proper ways to clean and store a paintbrush.

_____

_____

_____

_____

12. With _____, the finish is atomized into tiny droplets that flow from the gun nozzle, producing a flat, nearly flawless film.

_____

13. Identify the following equipment shown.

_____

Hitachi Power Tools U.S.A., Ltd.

14. _____ guns do not have a valve to shut off airflow.

_____

15. _____ guns allow the user to control the air and liquid flow rates.

_____

16. Identify the following types of nozzles.

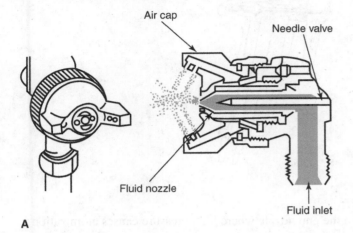

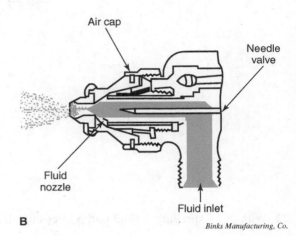

Binks Manufacturing, Co.

A. _____   B. _____

17. Using the Troubleshooting Guide for Air Spray Systems chart in the textbook, determine the probable cause(s) and remedy for a top-heavy spray pattern.

_____

_____

18. Using the Troubleshooting Guide for Air Spray Systems chart in the textbook, determine the probable cause and remedy for a split-spray pattern.

_____

19. Using the Troubleshooting Guide for Air Spray Systems chart in the textbook, determine the probable cause and remedy for no fluid flow.

_____

20. Explain how to use the wet film thickness gauge, and explain why it is important.

_____

_____

_____

_____

_____

_____

21. On the following spray gun, indicate where lubrication would be needed.

ITW DeVilbiss

22. With _____ spraying, a fluid pump forces finish up into the gun nozzle where high pressure causes atomization.

_____

23. List three advantages of airless equipment over air spray equipment.

_____

_____

_____

24. Identify the parts of the commercial airless spray system.

A. _____

B. _____

C. _____

D. _____

E. _____

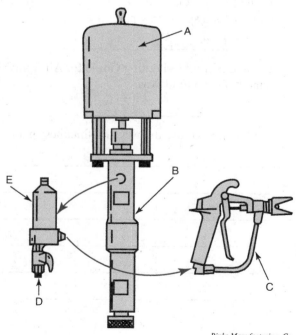

Binks Manufacturing, Co.

25. _____ are a handy alternative to using bulky and inconvenient spray equipment for finishing smaller products.

_____

26. _____ is a simple coating practice used to apply stain and penetrating finishes.

_____

27. _____ is a technique used to apply an even layer to small components quickly.

_____

28. List four examples of industrial finishing equipment.

_____

_____

_____

_____

29. List at least four safety precautions to take when applying coating materials.

_____

_____

_____

_____

◣ Activity 52A

Finish a Piece in Your Garage

Imagine you have just completed a small trinket box working with hand tools in your garage. In keeping with this same idea of simplicity, you want to finish this box using the simplest and safest methods. Explain what materials and tools you would need and how you would proceed with the process of applying a finish to your trinket box.

_____

_____

_____

_____

_____

_____

_____

_____

_____

_____

_____

_____

_____

_____

_____

_____

_____

_____

_____

_____

_____

_____

_____

_____

_____

_____

_____

_____

_____

_____

_____

_____

 Activity 52B

Set Up a Finishing Department

Imagine you are the owner of a growing successful mill and cabinet company. You have been sending all of your cabinets out to be stained and finished but now you would like to set up your own finishing department. In the space provided, draw a simple floor plan of your finished room and label the finishing equipment you would use and the locations of the specific finishing operations. Pay attention to product flow through the department. In the space following your drawing, describe your operation.

_____

_____

_____

_____

_____

_____

_____

_____

_____

_____

_____

_____

_____

_____

Notes

CHAPTER 53	Stains, Fillers, Sealers, and Decorative Finishes

Chapter Review

Carefully read Chapter 53 of the text and answer the following questions.

1. What does preparing a surface for topcoating involve?

   _____

   _____

   _____

   _____

   _____

2. A(n) _____ is a coat of thinned sealer that helps control stain penetration.

   _____

3. _____ alters a wood's color, accents grain patterns, or hides unattractive grain.

   _____

4. According to the **Stain Characteristics** chart in the textbook, what are two advantages of using penetrating oil as a stain?

   _____

5. It is wise to stain a(n) _____ or hidden area on a cabinet first.

   _____

6. _____ *True or False?* Water stain tends to raise the wood grain.

7. What is one disadvantage of non-grainraising (NGR) stain?

   _____

   _____

8. _____ *True or False?* Oil stains penetrate more deeply than water stains.

9. _____ stains are used in furniture restoration and for evening the tone between sapwood and heartwood.

   _____

10. _____ *True or False?* Latex stain is recommended for high-quality cabinetmaking.

11. _____ stain is similar to latex or pigment stains in that it hides the surface.

   _____

12. List six tips to follow when staining.

_____

_____

_____

_____

_____

_____

13. _____ is the process of packing a paste material into large pores of open grain woods, leveling the surface in preparation for topcoating.

_____

14. The most common filler is _____.

_____

15. _____ *True or False?* Filler can be applied before, after, or along with stain.

16. _____ *True or False?* Large pores in wood require a thinner filler mix than do small pores.

17. List the 11 steps involved in applying filler.

_____

_____

_____

_____

_____

_____

_____

_____

_____

_____

_____

18. Give two reasons sealer is needed on wood surfaces.

_____

_____

19. List four types of sealers.

_____

_____

_____

_____

20. _____ Which method provides the best results when applying sealers to wood?
    A. Brushing
    B. Rubbing
    C. Spraying
    D. Dipping

21. The sealing coat on metal is usually thinned clear _____ used to help the topcoat adhere better.

_____

22. Priming materials must be compatible with the _____.

_____

23. When priming metal products, apply a colored primer that will _____ with the topcoat color.

_____

24. List eight decorative finishes that can be applied to cabinetry and furniture.

_____

_____

_____

_____

_____

_____

_____

_____

25. _____ Which of the following involves outlining and highlighting edges and other areas with gold paint?
    A. Antiquing
    B. Gilding
    C. Shading
    D. Graining

26. List three safety rules for applying stains, fillers, sealers, and decorative finishes.

_____

_____

_____

 Activity 53A

Choose a Decorative Finish

After reading the textbook chapter, choose the decorative finish you found the most interesting. What kind of project would you apply it to? Explain what materials you would need and the process you would follow to apply the finish to your project.

_____

_____

_____

_____

_____

_____

_____

_____

_____

_____

_____

_____

_____

_____

_____

_____

_____

_____

_____

_____

_____

_____

_____

_____

_____

_____

_____

_____

CHAPTER 54 Topcoatings

Chapter Review

Carefully read Chapter 54 of the text and answer the following questions.

1. _____ is the final protective film on a completed product.

   _____

2. _____ coatings hide the surface and are applied to manufactured wood products and poor-quality materials.

   _____

3. Name three environmental factors that must be considered when applying a topcoating.

   _____

   _____

   _____

4. What is the primary advantage of penetrating finishes over built-up films?

   _____

   _____

5. _____ *True or False?* Raw linseed oil requires more drying time than boiled linseed oil.

6. List the six steps involved in applying a linseed oil finish.

   _____

   _____

   _____

   _____

   _____

   _____

7. List the seven steps involved in applying a tung oil finish.

   _____

   _____

   _____

   _____

   _____

   _____

   _____

8. _____ is a combination of oil and wax, and possibly stain.

   _____

9. _____ *True or False?* A phenolic resin finish is more durable than an alkyd resin coating.

10. _____ Which of the following is a low-build coating?
    A. Varnish
    B. Shellac
    C. Enamel
    D. Polyurethane

11. Why should you check the expiration date on shellac container labels?

    _____

    _____

12. The _____ process consists of applying a combination of shellac blended with other ingredients and oils.

    _____

13. _____ The _____ in synthetic lacquer gives the dried film added flexibility.
    A. resin
    B. plasticizer
    C. solvent
    D. None of the above.

14. _____ *True or False?* Varnishes sold under the names *spar* or *marine* are extra-tough products.

15. In _____ varnish, the resin used is rosin.

    _____

16. _____ Which of the following is the most wear-resistant of synthetic varnishes?
    A. Acrylic varnish
    B. Epoxy varnish
    C. Polyurethane varnish
    D. Phenolic varnish

17. _____ *True or False?* It is best to shake varnish to mix the ingredients.

18. At the proper viscosity, varnish spreads easily by _____.

    _____

19. _____ *True or False?* Most enamels today are natural.

20. _____ enamel is the most popular opaque finish for interior and exterior applications.

    _____

21. List eight characteristics of synthetic water-based enamels.

    _____

    _____

    _____

    _____

    _____

    _____

    _____

    _____

22. _____ Apply enamel as you would apply _____.
    A. varnish
    B. shellac
    C. lacquer
    D. None of the above.

23. Most built-up finishes require _____ or more coats of finish.

_____

24. Between coats, a deglossed surface should be rubbed with a(n) _____ grit or finer abrasive.

_____

25. The gloss of shellac, lacquer, and varnish can be changed by _____.

_____

26. What are two ways of making a nonscratch surface?

_____

27. Liquid removers are most appropriate for _____ surfaces.

_____

28. Using the **How to Remedy Surface Defects** chart in the textbook, describe the probable cause and remedy for pinholing a cabinet surface.

_____

_____

_____

29. Why should paste remover be brushed in only one direction?

_____

30. List five safety precautions to take when working with finishes and solvents.

_____

_____

_____

_____

_____

 **Activity 54A**

Choose a Topcoating

The following lists several pieces. Describe what topcoating you would use for each. Explain in detail why you chose that topcoating, and contrast your decision with the topcoatings you did not choose.

1. An Adirondack chair made of cedar that is placed outside all year round in direct sunlight.

_____

_____

_____

_____

_____

_____

_____

2. An island made of ash installed in a kitchen.

_____

_____

_____

_____

_____

_____

_____

3. A paint-grade cabinet made of MDF.

_____

_____

_____

_____

_____

_____

_____

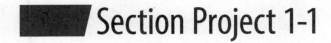 Section Project 1-1

Class PowerPoint® Presentation

Before beginning your study of cabinetmaking, it is important to understand the scope of the cabinetmaking industry. This project will help you build a foundation of knowledge about the industry.

Resource Chapters

Chapter 1—Introduction to Cabinetmaking
Chapter 6—Components of Design

Objectives

After completing this project, you should be able to:

- Understand the elements that make up the cabinetmaking industry.
- Understand how to use Microsoft PowerPoint to make a presentation.

Materials and Resources

- Computer with Microsoft PowerPoint
- Scanner

Activity Procedure

1. Read an assigned section of the chapter as an individual or in a team.

2. Scan pictures from the textbook and other sources.

3. Create PowerPoint slides.

4. Put your name on each slide you create. The slides of your entire class will be combined into one PowerPoint presentation.

5. Present the PowerPoint as a class, and present your specific section.

Final Project Requirements

- Your PowerPoint slides
- Class presentation
- Follow-up questions

Follow-Up Questions

1. What were your greatest challenges in completing this activity?

_____

_____

_____

2. What main idea stands out to you in this activity?

_____

_____

_____

3. How do you feel or think you did on your presentation?

_____

_____

_____

Rubric for Class PowerPoint Presentation

Student Guidelines

Oral presentation
Possible points _____ Score _____

Did you speak with confidence?

_____

Slide readability
Possible points _____ Score _____

Is your main idea obvious? Is the mix of text and images appropriate?

_____

Information accuracy
Possible points _____ Score _____

Is your information accurate?

_____

Subject coverage
Possible points _____ Score _____

Is your subject coverage thorough?

_____

Follow-up questions
Possible points _____ Score _____

Were all questions answered thoughtfully?

_____

Total possible points _____ Score _____

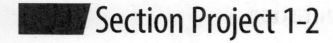 Section Project 1-2

Mini Flip Chart of Professional Organizations

Professional cabinetmaking organizations help their members stay up-to-date on the latest developments in the industry and techniques being used. Prepare a directory mini flip chart of the professional organizations that relate to cabinetmaking professionals.

Resource Chapters

Chapter 3—Career Opportunities
Chapter 4—Cabinetmaking Industry Overview

Objectives

After completing this project, you should be able to:

- Use the mini flip chart to organize information.
- Discuss at least six cabinetmaking organizations.

Materials and Resources

- Computer with a word processing program and internet access for conducting research on cabinetmaking organizations
- Four sheets of 8 1/2″ × 11″ paper

Activity Procedure

1. Conduct your research. Search using phrases such as "cabinetmaking professional organizations," "cabinetmaking supply organizations," and "architectural millwork organizations."

2. Collect four sheets of paper. Measure and make a small mark on one end of each piece, 1/2″ from the bottom edge of the paper.

3. Slide the sheets so that just the 1/2″ is exposed.

4. Fold the papers so that you get eight 1/2″ flaps.

5. Staple twice at the fold point. Study the diagram provided.

6. On the flaps (the 1/2″ exposed area), write the organization names you are researching.

7. Describe the organization under the flap. Include contact information, websites, email addresses, and a brief synopsis of the services the organization performs.

Final Project Requirement

- Completed mini flip chart

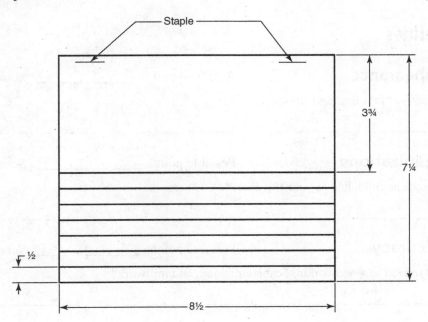

Goodheart-Willcox Publisher

Follow-Up Questions

1. What were your greatest challenges in completing this project?

   _____
   _____
   _____
   _____

2. Explain how this project will be used.

   _____
   _____
   _____
   _____

3. Which organization would you join first? Explain why.

   _____
   _____
   _____
   _____

Rubric for Mini Flip Chart of Professional Organizations

Student Guidelines

Format and appearance
Possible points _____ Score _____

Is your flip chart attractive? Is it neat and straight?

_____

Number of organizations
Possible points _____ Score _____

How many organizations are included in your flip chart?

_____

Information accuracy
Possible points _____ Score _____

Are all organizations related to woodworking, cabinetmaking, and millwork?

_____

Follow-up questions
Possible points _____ Score _____

Did you answer the questions thoughtfully?

_____

Total possible points _____ Score _____

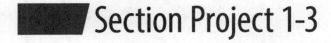 ## Section Project 1-3

Identifying Shop Hazards

Everyone should contribute to maintaining a safe atmosphere in the shop. This activity gives you a chance to be a part of the safety program in your shop. Identify any hazards you see in the shop and prepare a sign warning others of the hazards.

Resource Chapters

Chapter 2—Health and Safety
Chapter 6—Components of Design

Objectives

After completing this project, you should be able to:

- Communicate about hazards in the shop.
- Provide a detailed explanation of your chosen safety rule.
- Demonstrate use of the basic elements of design.
- Demonstrate resourcefulness and creativity.

Materials and Resources

- Graph paper
- Poster board
- Markers and colored pencils

Activity Procedure

1. Brainstorm ideas.

2. Decide on a final design and sketch it on a sheet of 8 1/2″ × 11″ graph paper.

3. Lay out your design in pencil on poster material.

4. When you are satisfied with the proportions of the lettering and the elements, darken and color.

5. Present your poster to your class.

6. Hang your poster in the shop.

Final Project Requirements

- Preliminary sketches
- Final poster design
- Follow-up questions

Follow-Up Questions

1. What were your greatest challenges in completing this project?

_____

_____

_____

2. Explain how this project will be used.

_____

_____

_____

_____

3. What safety concept did you decide to use for your poster?

_____

_____

_____

_____

4. Where did you hang your poster in the shop? Explain why.

_____

_____

_____

_____

Name _____

Rubric for Identifying Shop Hazards

Student Guidelines

Sketches
Possible points _____ Score _____

Did brainstorming lead to good choices for your design? Is your sketch a detailed, accurate description of the final poster design?

_____

Resourcefulness
Possible points _____ Score _____

Did you use time and materials wisely?

_____

Creativity and layout
Possible points _____ Score _____

Did you use materials creatively and follow the elements of design to communicate your idea?

_____

Visual communication
Possible points _____ Score _____

Is your main idea visible from across the room? Is 80% of remaining information visible from eight feet?

_____

Information accuracy
Possible points _____ Score _____

Is the information on the poster correct and complete?

_____

Craftsmanship
Possible points _____ Score _____

Did you follow the plans? Is your work neat and creative?

_____

Follow-up questions
Possible points _____ Score _____

Did you answer the questions thoughtfully?

_____

Total possible points _____ Score _____

Notes

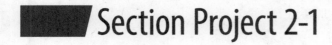

Section Project 2-1

Project Management

Almost all companies hire project managers. A project manager manages all the resources needed to complete a project. Project managers are the key to having materials and workers on a job at the right time and for the right price. An objective of this program is for you to develop project management skills. The documents in this activity are tools to help you manage all the projects in this workbook.

For this activity, you are going to apply the documents to a project either you or your instructor will choose. The best approach is to choose a project that does not have a drawing and complete the drawings and the planning for that project.

Resource Chapters

Chapter 9—Production Decisions
Chapter 10—Sketches, Mock-Ups, and Working Drawings
Chapter 11—Creating Working Drawings
Chapter 12—Measuring, Marking, and Laying Out Materials

Objectives

After completing this project, you should be able to:
- Describe the elements of project management.
- Make sketches of a project to scale using the graph paper provided in this project.
- Use a plan of procedure to manage the steps and your time in this project.
- Use the bill of materials sheet to figure the total cost of a project.
- Use the 4′ × 8′ layout plan and cut list form to plan the cutting process on 4′ × 8′ panel materials.
- Use the cut worksheet to create a material cut list.

Materials and Resources

- Project management sheets included in this activity

Activity Procedure

1. Study the completed project management sheets for the trinket box included in **Section Project 5-2** of this workbook.

2. Draw up the project using the graph paper sheets. Include appropriate pictorial sketches, multiview drawings, detail drawings, and section view drawings, complete with dimensions and notes. You should have a minimum of three pages in your drawing set. Make the drawings to scale by establishing a certain measurement to equal a square; for example, 1″ = 1 square. An object 3″ inches tall would be three squares tall on the graph paper. Your scale will mostly likely be different than 1″ = 1 square. Choose the scale that makes the object fit the paper best.

3. If sheet materials are to be used, use the 4′ × 8′ layout plan and cut worksheet to plan cuts.

4. Fill out the bill of materials sheet. This should include all materials, as if you were going to the supply store to get these materials. In most cases, you cannot buy exact sizes of sheet material. Therefore, your bill of materials will not have the exact sizes of panel goods on it, but will simply reference how many sheets are required.

5. Complete the plan of procedure. The form included in this project has space to keep track of time spent on each procedure. Shade in when each step starts and when it ends. This helps you, the project manager in this instance, plan when you are going to need resources.

Final Project Requirements

- Shop drawing of the project you designed
- Layout plans for the 4′ × 8′ sheets
- Bill of materials for the project
- Cut worksheet for solid materials (ask your instructor if you should include the 4′ × 8′ panel materials you placed on that sheet)
- Plan of procedure for the project

Name _____

NAME : _____

DRAWING SCALE ____ = ____ SQUARES

DATE ___/___/___   FILE No. _____

Name _____

NAME : _____   DRAWING SCALE ____ = ____ SQUARES   DATE ___/___/___   FILE No. _____

4' × 8' Layout Plan and Cut List

Name or Job _____ Period _____ Date _____

Part Label	Pieces	Thickness	Width	Length	Square Inches	Square Feet

Material Cut Sheet

Part Name	Material	Number Pieces	Rough Measurements			Cut Process Completed	Finish Measurements			Cut Process Completed
			Thickness	Width	Length		Thickness	Width	Length	

Bill of Materials

Project Name: _____

Quantity	Units	Material Description	Price/Unit	Extended Price
			Subtotal	
			Tax	
			Total	

Name _____

Plan of Procedure

Minutes, Hours, Days, Weeks, or Months

Steps	1	2	3	4	5	6	7	8	9	10	11	12	13	14	15	16	17	18	19	20	21	22	23	24	25
1																									
2																									
3																									
4																									
5																									
6																									
7																									
8																									
9																									
10																									
11																									
12																									
13																									
14																									
15																									
16																									
17																									
18																									
19																									
20																									
21																									
22																									
23																									
24																									
25																									
26																									
27																									
28																									
29																									
30																									
31																									
32																									

Follow-Up Questions

1. What were your greatest challenges in completing the process plan?

   _____

   _____

   _____

   _____

2. Explain how the plan of procedure was used.

   _____

   _____

   _____

   _____

3. Explain in your own words what is involved in project management.

   _____

   _____

   _____

   _____

4. If you leave out a step on the process plan and realize it later, what would you do to correct the plan?

   _____

   _____

   _____

   _____

5. Explain how you use graph paper to draw objects to scale.

   _____

   _____

   _____

   _____

Rubric for Project Management

Student Guidelines

Shop drawings
Possible points _____   Score _____

Are dimensions applied correctly and drawn to scale? Are all necessary details present?

_____

Bill of materials
Possible points _____   Score _____

Are materials figured correctly? Are all materials listed and calculated?

_____

Plan of procedure
Possible points _____   Score _____

Are essential steps listed in order? Are there time estimates? Are actual time records kept?

_____

4′ × 8′ layout plan and cut list
Possible points _____   Score _____

Was the form used to determine the number of sheets needed on the materials bill?

_____

Cut worksheet
Possible points _____   Score _____

Was the cut worksheet filled out in detail?

_____

Total possible points _____   Score _____

Name _____ Date _____ Class _____

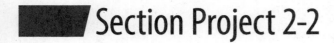

Student Desk and Desk Chair Design

In this project, you will design a simple student desk and desk chair. Choose a cabinetry style that you like. You will draw shop drawings, including pictorial and multiview drawings to scale on graph paper, showing overall dimensions. The human factors should guide all of your dimensions.

Resource Chapters

Chapter 5—Cabinetry Styles
Chapter 6—Components of Design
Chapter 7—Design Decisions
Chapter 8—Human Factors
Chapter 10—Sketches, Mock-Ups, and Working Drawings
Chapter 11—Creating Working Drawings

Objectives

After completing this project, you should be able to:

- Demonstrate an understanding of cabinetry styles by explaining what influenced your choice of style in the chosen design.
- Demonstrate an understanding of the components of design by explaining how each of the components were considered in the design.
- Demonstrate an understanding of the design decision process by documenting the process.
- Demonstrate an understanding of the human factors in the final design.
- Demonstrate a mastery of preliminary sketching for a design.
- Demonstrate an understanding of the different methods of pictorial sketching by creating an isometric, cabinet oblique, or perspective sketch of the chosen design.
- Demonstrate an understanding of a two-view multiview drawing by making a two-view shop drawing of the final design on graph paper.

Materials and Resources

- Textbook
- Graph paper
- Sharp pencils

Activity Procedure

1. Make rough sketches of your desk and chair. Decide exactly what you want. For instance, do you want drawers? Shelves?

2. Consider human factors in designs. Collect all the measurements you need, such as chair height and desk leg clearance.

3. Consider what style is going to influence your design.

4. Make several different thumbnail sketches of possible designs that meet the specifications you have chosen.

5. Choose and refine your design by making more detailed pictorial sketches.

6. Draw to scale two- or three-view multiview drawings on graph paper.

7. Make necessary detail drawings of parts and assemblies that cannot be easily seen in the multiview drawing.

Final Project Requirements

- A list of the human factors you considered in your design
- A statement of your design style choice
- A statement of how the components of design were used in the design of your products
- A statement explaining how the design process was used to arrive at the final design
- Preliminary sketches drawn on plain paper
- A refined pictorial sketch done on graph paper to scale of each item
- A two- or three-view drawing of each item
- At least one sketch on graph paper of a detail drawing and section drawing showing complicated construction detail

The Decision-Making Process

1. Identify needs and wants. Specify what problems this product must solve.

_____

_____

_____

_____

_____

_____

2. Gather information. What is the current style, if any, that should be matched? Who will be using the product? List the human factors.

_____

_____

_____

_____

_____

3. Create at least six preliminary ideas. Sketch each idea.

4. Pick three of your ideas as possible final designs, and refine them.

Idea 1

Idea 2

Idea 3

5. Analyze the function, strength, cost, and appearance of your ideas.

Positive aspects of Idea 1 _____

_____

_____

Negative aspects of Idea 1 _____

_____

_____

_____

Name _____

Positive aspects of Idea 2 _____

_____

_____

Negative aspects of Idea 2 _____

_____

_____

Positive aspects of Idea 3 _____

_____

_____

Negative aspects of Idea 3 _____

_____

_____

6. Make a decision. Choose your final idea based on the analysis you have just made. Consider the final style, drawer arrangement, doors, and proportions of each item. Explain your decision.

_____

_____

_____

_____

_____

7. On graph paper, draw detailed shop drawings needed to build the final design. Your drawing set should include a separate drawing of the desk and chair, and detail or section drawings of each drawer and door with at least one section and detail included in the drawing set.

Rubric for Student Desk and Desk Chair Design

Student Guidelines

Completeness
Possible points _____ Score _____

Are there the necessary number of drawings and views to fully explain the solution and decision?

_____

Style
Possible points _____ Score _____

Do the drawings represent a specific style?

_____

Dimensions
Possible points _____ Score _____

Are extension lines, dimension lines, and either tic marks, arrowheads, or dots to end dimension lines used properly?

_____

View accuracy
Possible points _____ Score _____

Do views on the same page align? Do they accurately represent the final design?

_____

Decision-making form
Possible points _____ Score _____

Does the work represent thoughtful planning? Does it follow the process?

_____

Total possible points _____ Score _____

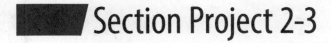 ## Section Project 2-3

Components of Design Poster

An understanding of the components of design is an important part of making desirable designs. Design a poster that communicates the four design principles of harmony, repetition, balance, and proportion. The poster design should follow the components of design.

Resource Chapter
Chapter 6—Components of Design

Objectives
After completing this project, you should be able to:
- Demonstrate an understanding of scale.
- Demonstrate use of the basic components of design as referenced in Chapter 6.
- Demonstrate resourcefulness and creativity.

Materials and Resources
- 8 1/2″ × 11″ graph paper
- 18″ × 24″ poster board
- Miscellaneous poster and graphic materials, such as construction paper and markers

Activity Procedure

1. Sketch ideas on a piece of 8 1/2″ × 11″ graph paper.

2. Be detailed in this planning stage. Include decisions on colors and materials. For instance, if a purple construction paper cutout is to be included on the poster, label that on the sketch. Someone else should be able to look at your sketch and create the poster.

3. Decide on size and boldness of your print. Remember, the idea here is to communicate clearly. The final poster will be twice the size of the sketch, so all elements on the sketch will be drawn half size.

4. Create the final poster on 18″ × 24″ poster board.

Final Project Requirements
- Preliminary scale sketches on 8 1/2″ × 11″ graph paper
- Final poster

Follow-Up Questions

1. What were your greatest challenges in completing this project?

_____

_____

_____

2. Explain how this project will be used.

_____

_____

_____

3. What element of the design concept do you feel you understand the best? Please explain.

_____

_____

_____

Name _____

Rubric for Components of Design Poster

Student Guidelines

Sketch
Possible points _____ Score _____

Was the sketch a detailed, accurate description and graphic representation of the final poster?

_____

Resourcefulness
Possible points _____ Score _____

Did the student use all materials available in a useful manner?

_____

Creativity and layout
Possible points _____ Score _____

Did creative use of materials and the elements of design accomplish the target goal of communicating an idea?

_____

Accuracy of information
Possible points _____ Score _____

Is information on the poster correct and complete?

_____

Craftsmanship and neatness
Possible points _____ Score _____

Was the design well executed?

_____

Visual communication
Possible points _____ Score _____

Is the main idea visible from across the room? Is 80% of remaining information visible from eight feet?

_____

Total possible points _____ Score _____

Notes

_____

_____

_____

_____

_____

_____

_____

_____

_____

_____

_____

_____

_____

_____

_____

_____

_____

_____

_____

_____

_____

_____

_____

_____

_____

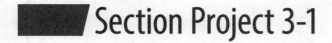 Section Project 3-1

Wood Samples Stringer

Having a sample of all the materials that are available in your shop can be very helpful when designing pieces or deciding what material to choose for your next project.

Resource Chapters

Chapter 15—Cabinet and Furniture Woods
Chapter 16—Manufactured Panel Products
Chapter 23—Sawing with Stationary Power Machines

Objectives

After completing this project, you should be able to:

- Explain the properties of all the materials used in the cabinet shop.
- Explain what units the materials are sold in (sheets, board foot, square foot, etc.).
- List the materials used in your cabinetmaking shop.

Materials and Resources

- A 2″ × 5″ sample of each type of material used in your shop
- Computer with word processing software (optional)
- Laser printer (optional)

> **SAFETY NOTE**
>
> Before proceeding with this activity, you should have passed a general shop safety rule test, all specific safety tests on the tools used in this activity, and be certified by the instructor to use the tools and equipment needed for this activity.

Activity Procedure

1. Work with your classmates to cut solid wood material, plywood, manufactured panel materials, plastic laminate, and plastic acrylic sheets into 2″ × 5″ sample pieces.

2. Make labels to attach to each sample that include the following information:

    Weight per unit

    Price per unit

    Description of properties of the material

    Standard uses of the material

    Unit measure for the material (such as board foot, sheet, or piece)

3. Drill a 1/4″ hole in one end of each piece, 1/2″ from the end and 1″ from the side.

4. Tie pieces together using a piece of 1/8″ braided nylon rope. String the rope through the holes and tie a knot, forming a loop. Use enough rope so that you can easily flip through your samples.

Final Project Requirements

- Sample stringer with the number of labeled samples assigned by your instructor

Follow-Up Questions

1. What were your greatest challenges in completing this project?

   _____

   _____

   _____

   _____

2. Explain how this project will be used.

   _____

   _____

   _____

   _____

3. Did you finish your project on time?

   _____

4. What piece was your favorite piece or interested you the most? Explain why.

   _____

   _____

   _____

   _____

5. How many samples did you create?

   _____

Name _____

Rubric for Wood Samples Stringer

Student Guidelines

Description label
Possible points _____ Score _____

Were descriptions of units, weight per unit, and material uses adequate?

_____

Number of samples
Possible points _____ Score _____

Did you complete the assigned number of samples?

_____

Craftsmanship
Possible points _____ Score _____

Are mill marks sanded out, corners broken properly, and holes drilled consistently in each piece?

_____

Efficiency
Possible points _____ Score _____

Was the assignment finished on time?

_____

Follow-up questions
Possible points _____ Score _____

Were all questions answered in a thoughtful manner?

_____

Total possible points _____ Score _____

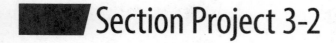 Section Project 3-2

Moisture Content Material Testing

Moisture content of wood is a serious concern to a cabinetmaker. In this project, you will determine the moisture content of five wood samples.

Resource Chapter

Chapter 13—Wood Characteristics

Objectives

After completing this project, you should be able to:
- Identify and describe open grain wood and closed grain wood.
- Differentiate between radially (quarter sawn) and tangentially (plain sawn) samples.
- Explain how to determine moisture content.
- Use technical terms from the chapter to describe the samples.

Materials and Resources

- Digital scale
- Magnifying glass
- Digital calipers
- Five different wood species samples, one cut radially, another tangentially
- Oven

> **SAFETY NOTE**
>
> Before proceeding with this activity, you should have passed a general shop safety rule test and all specific safety tests on the tools used in this activity, and be certified by the instructor to use the tools and equipment needed for this activity.

Activity Procedure

1. Cut one 3/4″ × 2″ × 4″ piece from each of the different wood species that are quarter sawn.

2. Cut one 3/4″ × 2″ × 4″ piece from each of the different wood species that are plain sawn.

3. Review the categories on the Lab Sheet and enter your findings.

4. Using an oven and following the instruction in the chapter, dry the wood samples to zero moisture content.

5. Use a magnifying glass to inspect the grain. Determine if it is open or closed grain.

   Note: Shaving the end grain to a smooth surface with a sharp knife can yield interesting results when viewed through a magnifying glass. This can also help you determine if the grain is open or closed. Small pores indicate closed grain; large pores indicate open grain.

Final Project Requirements

- All wood samples
- Completed Wood Properties Lab Sheet

Wood Properties Lab Sheet

	Radially (quarter sawn)					Tangentially (plain sawn)				
	Sample 1	Sample 2	Sample 3	Sample 4	Sample 5	Sample 6	Sample 7	Sample 8	Sample 9	Sample 10
Species										
Wet thickness										
Wet width										
Wet length										
Wet weight										
Dry weight										
Percent										
Moisture Content										
Dry thickness										
Difference wet thickness – dry thickness										
Dry width										
Difference wet width – dry width										
Dry length										
Difference Wet Length – Dry Length										
Open or closed grain										
Color										

Follow-Up Questions

1. What were your greatest challenges in completing this project?

_____

_____

_____

_____

2. Is it good or bad for wood to be at zero moisture content before using it in a cabinetmaking project? Explain your answer.

_____

_____

_____

_____

3. Did thickness, width, or length change the most after drying? Was this as predicted in the text?

_____

_____

_____

_____

4. What species seemed to be the densest and, therefore, the strongest?

_____

Name _____

Rubric for Moisture Content Material Testing

Student Guidelines

Completeness of lab sheet
Have you completed all sections of the Lab Sheet?

Possible points _____ Score _____

_____

Accuracy of findings
Are your findings comparable to what is expected?

Possible points _____ Score _____

_____

Reporting quality
Do you think you have met the objectives?

Possible points _____ Score _____

_____

Efficiency
Did you finish the work in the allotted time?

Possible points _____ Score _____

_____

Follow-up questions
Were all questions answered in a thoughtful manner?

Possible points _____ Score _____

_____

Total possible points _____ Score _____

Notes

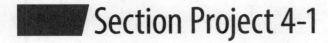

Section Project 4-1

Making a Bench Hook

When using hand tools, it can be a challenge to hold the stock while it is being worked. Clamps and vises can be inefficient or get in the way of the work. A bench hook or "shooting board" can be used in some of these situations.

Resource Chapters

Chapter 22—Sawing with Hand and Portable Power Tools
Chapter 23—Sawing with Stationary Power Machines
Chapter 24—Surfacing with Hand and Portable Power Tools
Chapter 27—Drilling and Boring

Objectives

After completing this project, you should be able to:

- Use the plans to determine materials needed.
- Use the cabinet workshop to complete a small project.
- Crosscut a small piece of stock to length using a fixture (bench hook) to hold the material.

Materials and Resources

Items you might need to complete this project include:

- 3/4″ thick particleboard
- Square
- Table saw
- Wood glue
- Drill press
- 1 1/2″ no. 10 screw
- Drill
- 1/8″ drill bit
- 3/16″ drill bit

SAFETY NOTE

Before proceeding with this activity, you should have passed a general shop safety rule test, all specific safety tests on the tools used in this activity, and be certified by the instructor to use the tools and equipment needed for this activity.

Activity Procedure

1. Study the bench hook plans shown here.

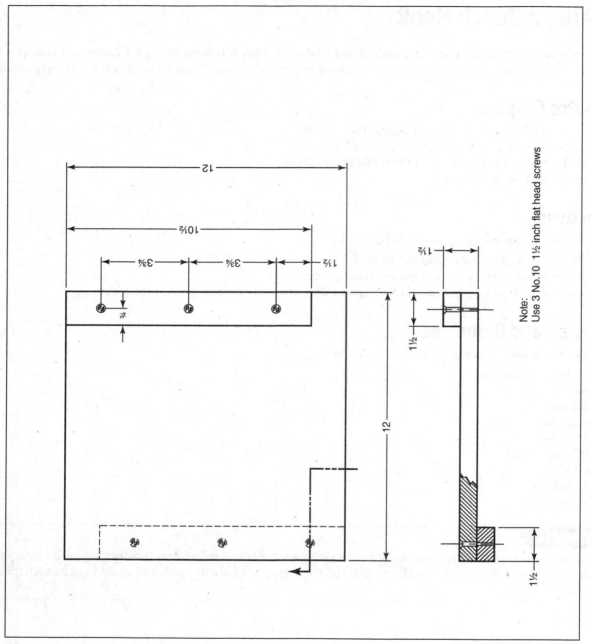

2. Make a cut list.

3. Make a bill of materials.

4. Make a list of resources needed such as tools and equipment.

5. Make a plan of procedure. Remember, this is the most important part of getting ready to work. You should be able to follow your plan step-by-step while executing your project.

6. Execute your plan of procedure. Refer to pages 106 and 107 in the text for examples of procedural steps.

Name _____

Final Project Requirements

- Bill of materials
- Plan of procedure
- A completed bench hook

- Material cut sheet
- Follow-up questions

Bill of Materials

Project Name: _____

Quantity	Units	Material Description	Price/Unit	Extended Price
			Subtotal	
			Tax	
			Total	

Material Cut Sheet

Part Name	Material	Number Pieces	Rough Measurements			Cut Process Completed	Finish Measurements			Cut Process Completed
			Thickness	Width	Length		Thickness	Width	Length	

Name _____

Plan of Procedure

Minutes, Hours, Days, Weeks, or Months

Steps	1	2	3	4	5	6	7	8	9	10	11	12	13	14	15	16	17	18	19	20	21	22	23	24	25
1																									
2																									
3																									
4																									
5																									
6																									
7																									
8																									
9																									
10																									
11																									
12																									
13																									
14																									
15																									
16																									
17																									
18																									
19																									
20																									
21																									
22																									
23																									
24																									
25																									
26																									
27																									
28																									
29																									
30																									
31																									
32																									

Follow-Up Questions

1. What were your greatest challenges in completing this project?

   _____

   _____

   _____

   _____

2. Explain how this project will be used.

   _____

   _____

   _____

   _____

3. Did you finish your project on time? What processes went faster than you thought they would and what processes took longer?

   _____

   _____

   _____

   _____

Rubric for Making a Bench Hook

Student Guidelines

Materials and resources
Possible points _____ Score _____

Did you supply a complete bill of materials, complete list of resources, and total cost of materials calculated?

_____

Plan of procedure

Did you create a detailed list of steps to build the project, and keep accurate time records of estimated time and actual time for each step?

_____

Craftsmanship

Did you follow plans? Are corners broken properly, edges flush at joints, and screw heads countersunk?

_____

Follow-up questions

Were all questions answered in a thoughtful manner?

_____

Total possible points _____ Score _____

 Section Project 4-2

Drill Press Practice

This activity will provide the opportunity for practice on the drill press.

Resource Chapters

Chapter 22—Sawing with Hand and Portable Power Tools
Chapter 23—Sawing with Stationary Power Machines
Chapter 24—Surfacing with Hand and Portable Power Tools
Chapter 27—Drilling and Boring
Chapter 37—Joinery

Objectives

After completing this project, you should be able to:

- Read a view drawing.
- Set the drill press to a certain depth.
- Change drill bits in a drill press.

- Lay out location of holes.
- Drill holes accurately to a layout.

Materials and Resources

- 2 × 4 pine or similar material
- Fully functional drill press

Activity Procedure

1. Cut a piece of 2 × 4 material to 6″ long. A 2 × 4 is 1 1/2″ × 3 1/2″ actual dimensions.

2. Use a square and pencil to mark the location of the hole on the material.

3. Install the 3/8″ twist drill.

4. Set the drill press depth stop so that the 3/8″ holes will drill down 3/4″.

5. Bore four 3/8″ holes in the corners.

6. Install a 3/4″ sharp spade bit.

7. Install a drill press vise on the drill press

8. Place a sacrificial board in the vise that is slightly smaller than the workpiece. The sacrificial board will keep the wood from splintering through on the back side and also give the spade bit point something to drill into.

9. Drill the hole.

10. Install a 1/4″ drill bit in the drill press.

11. Turn the piece up on edge, reclamp in the vise, and drill the 1/4″ hole as per plans.

12. Turn the workpiece up on end, reclamp in the vise, and drill the second 1/4″ hole as per plans.

13. Submit your completed project.

Final Project Requirements

- Completed project
- Follow-up questions

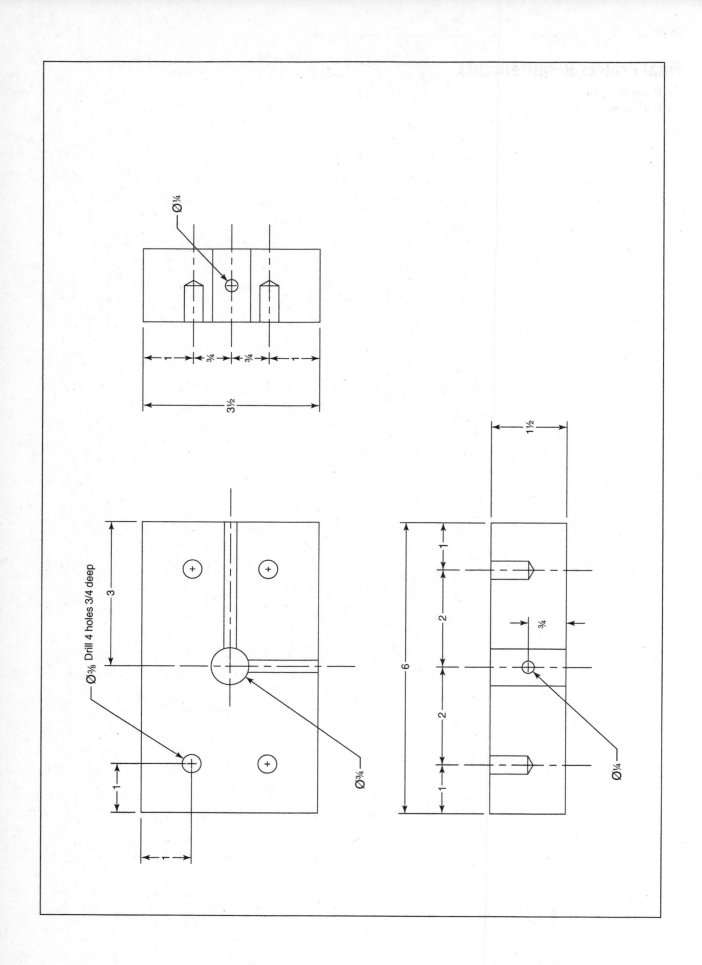

Follow-Up Questions

1. What were your greatest challenges in completing this project?

   _____

   _____

   _____

   _____

2. Did you finish your project on time? What processes went faster than you thought they would and what processes took longer?

   _____

   _____

   _____

   _____

   _____

3. Explain how you kept from drilling the four 3/8″ diameter holes too deep.

   _____

   _____

   _____

   _____

4. How did you hold the work while drilling to keep it steady?

   _____

   _____

   _____

   _____

5. What kind of bit did you use to drill the 3/4″ hole?

   _____

   _____

Rubric for Drill Press Practice

Student Guidelines

Plan of procedure
Possible points _____ Score _____

Did you provide a detailed list of steps to build the project, and keep accurate time records of estimated time and actual time for each step?

_____

Craftsmanship
Possible points _____ Score _____

Did you follow plans, drill holes in proper locations, and follow proper procedure on all operations?

_____

Follow-up questions
Possible points _____ Score _____

Did you answer all questions in a thoughtful manner?

_____

Total possible points _____ Score _____

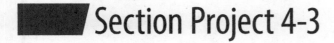 Section Project 4-3

Dry Fit

Each project you make will provide unique challenges in clamping. Most cabinetmakers will dry fit all joints before adding glue. In this step, they will also determine how to apply clamping pressure. Glued joints without pressure do not have the strength of glued joints that have had pressure applied.

Resource Chapter

Chapter 32—Gluing and Clamping

Objective

After completing this project, you should be able to:

- Be able to use all clamps available in your shop to clamp materials in various shapes and sizes.

Materials and Resources

- Variety of bar clamps
- Hand-screw clamps
- C-clamps
- Miscellaneous specialty clamps available in your shop
- Vacuum press
- Variety of wood pieces, cut square and with miters (edges and end should be true, as if they were going to be glued)

Activity Procedure

> **SAFETY NOTE**
>
> Before proceeding with this activity, you should have passed a general shop safety rule test, all specific safety tests on the tools used in this activity, and be certified by the instructor to use the tools and equipment.

1. Find two pieces of wood, approximately 6″ each.

2. Clamp the wood pieces face-to-face using hand-screw clamps. Practice placing the jaws parallel in such a way that pressure is applied evenly across the entire jaw of the clamp. Ask your instructor to observe your setup and mark the rubric section.

3. Find five pieces of wood that are approximately 3/4″ × 3″ × 18″.

4. Using bar clamps, clamp these pieces together as if they are to be joined edge-to-edge, forming a larger piece. Ask your instructor to observe your setup and mark the rubric section.

5. Find two small pieces of wood in the 4″–6″ size range.

6. Clamp the wood pieces with C-clamps. The two surfaces should be clamped solidly. Ask your instructor to observe your setup and mark the rubric section.

7. Set up the vacuum bag press. Place two pieces of wood face-to-face and put them in the press. Pull a vacuum on the pieces. Ask your instructor to observe your setup and mark the rubric section.

8. Find a piece of plywood and 1/4″ thick edgeband strips.

9. Use edge clamps to dry fit edge trim to the plywood. Ask your instructor to observe your setup and mark the rubric section.

Final Project Requirement

- Follow-up questions

Follow-Up Questions

1. What were your greatest challenges in completing this activity?

_____

_____

_____

_____

2. Explain how this activity experience will help you with later projects.

_____

_____

_____

_____

3. Explain how you were able to get even pressure with the hand-screw clamps.

_____

_____

_____

_____

Rubric for Dry Fit

Student Guidelines

Hand-screw clamps   Possible points _____   Score _____
Were jaws clamped parallel with even pressure? Was the correct number of clamps used for the material?

_____

Bar clamps   Possible points _____   Score _____
Was the correct number of clamps used for the material? Was the grain alternated?

_____

C-clamps   Possible points _____   Score _____
Were the pieces clamped face-to-face with the proper number of clamps? Were the edges flush?

_____

Vacuum bag press   Possible points _____   Score _____
Was the press set up properly and not leaking?

_____

Edge clamps   Possible points _____   Score _____
Were clamps set up properly?

_____

Follow-up questions   Possible points _____   Score _____
Did you answer all questions in a thoughtful manner?

_____

Total possible points _____   Score _____

 **Section Project 4-4**

Building a Machine Jig or Fixture

Machine jigs or fixtures can add production efficiency to your work. In this activity, you are going to develop a jig or fixture to make a common procedure more efficient.

Resource Chapters

Chapter 22—Sawing with Hand and Portable Power Tools
Chapter 23—Sawing with Stationary Power Machines
Chapter 24—Surfacing with Hand and Portable Power Tools
Chapter 27—Drilling and Boring
Chapter 37—Joinery
Chapter 38—Accessories, Jigs, and Special Machines

Objectives

After completing this project, you should be able to:

- Explain why shop-made accessories, jigs, and fixtures are important for shop production.
- List the most common accessories, jigs, and fixtures.
- Explain how a jig works.

Materials and Resources

- Scrap stock
- Plywood
- MDF
- Screws
- Dowels

Activity Procedure

> **SAFETY NOTE**
>
> Before proceeding with this activity, you should have passed a general shop safety rule test, all specific safety tests on the tools used in this activity, and be certified by the instructor to use the tools and equipment.

1. Decide which jig or fixture you want to make. Trade journals are a good source of ideas. Other ideas include:

   table saw—taper jig, both adjustable and nonadjustable, miter jig, splined plain miter jig, splined flat miter jig, tenon jig, outfeed table

   band saw—circle cutting fixture

   drill press—V-block, dowel pin drill fence

   picture frame clamp

   radial arm saw—push board

   band saw—auxiliary fence

2. Prepare a sketch of the fixture or jig on graph paper.

3. Prepare a plan of procedure.

4. Prepare a bill of materials.

5. Prepare a cut list.

Final Project Requirements

- Shop sketch on graph paper
- Bill of materials
- Cut list
- Plan of procedure
- Final fixture
- Demonstration of how your jig or fixture works
- Follow-up questions

DRAWING SCALE _____ = _____ SQUARES   DATE___/___/___   FILE No. _____

NAME : _____

Name _____

Plan of Procedure

Minutes, Hours, Days, Weeks, or Months

Steps	1	2	3	4	5	6	7	8	9	10	11	12	13	14	15	16	17	18	19	20	21	22	23	24	25
1																									
2																									
3																									
4																									
5																									
6																									
7																									
8																									
9																									
10																									
11																									
12																									
13																									
14																									
15																									
16																									
17																									
18																									
19																									
20																									
21																									
22																									
23																									
24																									
25																									
26																									
27																									
28																									
29																									
30																									
31																									
32																									

Bill of Materials

Project Name: _____

Quantity	Units	Material Description	Price/Unit	Extended Price
			Subtotal	
			Tax	
			Total	

Material Cut Sheet

Part Name	Material	Number Pieces	Rough Measurements			Cut Process Completed	Finish Measurements			Cut Process Completed
			Thickness	Width	Length		Thickness	Width	Length	

Follow-Up Questions

1. What were your greatest challenges in completing this project?

   _____

   _____

   _____

   _____

2. Explain how this project will be used.

   _____

   _____

   _____

   _____

3. What accessory, jig, or fixture did you construct?

   _____

   _____

   _____

   _____

4. Explain the difference between a jig and a fixture.

   _____

   _____

   _____

   _____

Rubric for Building a Machine Jig or Fixture

Student Guidelines

Print
Possible points _____   Score _____

Assess your project neatness, accuracy, dimensions, and correctness of views.

_____

Plan of procedure
Possible points _____   Score _____

Did you have a detailed list of steps to build the project, and keep accurate time records of estimated time and actual time for each step?

_____

Demonstration
Possible points _____   Score _____

Were you confident and organized while demonstrating your project?

_____

Craftsmanship
Possible points _____   Score _____

Did you follow the plans? Did joint represent perfect location of holes and cuts?

_____

Follow-up questions
Possible points _____   Score _____

Did you answer all questions in a thoughtful manner?

_____

Total possible points _____   Score _____

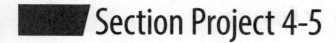 **Section Project 4-5**

Wood Joints Class Project

A thorough knowledge of joinery techniques is required to be considered a master cabinetmaker. In this activity, each student will be assigned one or two joints shown in the textbook chapter and make a sample of each.

Resource Chapters

Chapter 22—Sawing with Hand and Portable Power Tools
Chapter 23—Sawing with Stationary Power Machines
Chapter 24—Surfacing with Hand and Portable Power Tools
Chapter 27—Drilling and Boring
Chapter 37—Joinery

Objectives

After completing this project, you should be able to:

- Explain the technique to make at least one wood joint.
- Identify the joints in the activity.
- Identify the difference in non-positioned, positioned, and reinforced joints.

Materials and Resources

- Scrap wood
- Assorted finish nails
- Assorted screws
- 3/8″ dowels
- No. 20 biscuits
- Pocket screws
- Plate joiner
- Dowel jig
- Tenon jig
- Pocket screw jig or machine
- 4′ × 4′ board for mounting joints

SAFETY NOTE

Before proceeding with this activity, you should have passed a general shop safety rule test, all specific safety tests on the tools used in this activity, and be certified by the instructor to use the tools and equipment needed for this activity.

Activity Procedure

1. With your classmates' help, cut the scrap wood into 3″ × 6″ pieces.

2. Collect two pieces of wood to build your joint. Joints to consider include:

Box joint
Dovetail joint
    Handmade through single
    Dovetail joint, machine-made through multiple
    Dovetail joint, half-blind multiple
Face-to-face butt joint
Edge-to-edge butt joint
End-to-end butt joint
    Reinforced with dowels
End-to-edge butt joint
    Reinforced with pocket screws
End-to-face butt joint
    Reinforced with screws
    Reinforced with nails
    Reinforced with dowels
    Reinforced with glue block
    Reinforced with RTA fasteners
Dado joint
    Blind dado
    Half dado

Groove joint
Rabbet joint
    Half rabbet
    Dado and rabbet
    Dado tongue and rabbet
Lap joints
    T-lap
    End lap
    Middle lap
Miter joints
    Flat miter
    Plain miter
    Rabbet miter
    Half lap miter
    Lap and miter
Mortise and tenon
    Blind mortise and tenon
    Through mortise and tenon
    Open mortise and through tenon
Butterfly joint

3. Refer to the text to learn how to make your joint.

4. Sketch your joint on graph paper.

5. Make a plan of procedure.

6. Make your joint.

7. Prepare a five-minute oral report about the joint you made.

Final Project Requirements

- Sample assigned joint
- Report outline
- Follow-up questions

Name _____

Plan of Procedure

Minutes, Hours, Days, Weeks, or Months

Steps	1	2	3	4	5	6	7	8	9	10	11	12	13	14	15	16	17	18	19	20	21	22	23	24	25
1																									
2																									
3																									
4																									
5																									
6																									
7																									
8																									
9																									
10																									
11																									
12																									
13																									
14																									
15																									
16																									
17																									
18																									
19																									
20																									
21																									
22																									
23																									
24																									
25																									
26																									
27																									
28																									
29																									
30																									
31																									
32																									

Follow-Up Questions

1. What were your greatest challenges in completing this project?

_____

_____

_____

_____

2. Explain how this project will be used.

_____

_____

_____

_____

3. What joints did you complete?

_____

_____

_____

_____

Rubric for Wood Joints Class Project

Student Guidelines

Print
Possible points _____ Score _____

Assess neatness, accuracy, dimensions, correctness of views, and accuracy of plan of procedure.

_____

Plan of procedure
Possible points _____ Score _____

Did you have a detailed list of steps to build the project, and keep accurate time records of estimated time and actual time for each step?

_____

Oral presentation
Possible points _____ Score _____

Were you confident and organized in presenting information?

_____

Craftsmanship
Possible points _____ Score _____

Did you follow the plans? Did joints represent perfect location of holes and cuts?

_____

Follow-up questions
Possible points _____ Score _____

Were all questions answered in a thoughtful manner?

_____

Total possible points _____ Score _____

Name _____ Date _____ Class _____

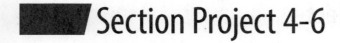 Section Project 4-6

Make a Push Stick from a Pattern

Using push sticks to keep fingers a safe distance from machinery saw blades is a habit all cabinetmakers should create. In this activity, you will make several push sticks using a pattern. It is a good idea to have multiple push sticks on hand because one is used as a sacrificial stick when cutting material narrower than 3/4″ wide.

Resource Chapters

Chapter 12—Measuring, Marking, and Laying Out Materials
Chapter 22—Sawing with Hand and Portable Power Tools
Chapter 23—Sawing with Stationary Power Machines

Objectives

After completing this project, you should be able to:

- Transfer a design from a square grid layout to a pattern.
- Use the band saw or jig saw to cut out an irregular shape.

Materials and Resources

- 8 1/2″ × 11″ card stock paper
- 8 1/2″ × 11″ plywood or solid wood
- 1/8″ or 1/4″ tempered hardboard for an original pattern
- Portable jig saw and band saw

> **SAFETY NOTE**
>
> Before proceeding with this activity, you should have passed a general shop safety rule test, all specific safety tests on the tools used in this activity, and be certified by the instructor to use the tools and equipment needed for this activity.

Activity Procedure

1. Find a piece of 8 1/2″ × 11″ card stock and mark off a 1/2″ grid pattern.

2. Using the activity sheet provided, lay out the pattern as explained in Chapter 12.

3. Glue this pattern to the piece of hardboard.

4. Using the band saw, cut out the pattern.

5. Using a disk sander, sandpaper, or files, smooth the edges, making them true to the pattern drawing.

6. Use the pattern to lay out two more push sticks. Retain this pattern for future use.

7. Cut one push stick using the band saw and following the principles explained in Chapter 23.

8. Cut another push stick using the jig saw and following the principles explained in Chapter 22.

Final Project Requirements

- Hardboard pattern
- Two 3/4″ plywood or solid wood push sticks as per drawing

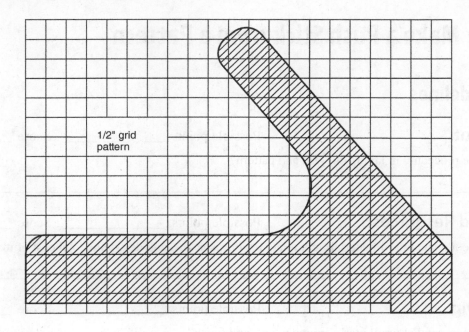

1/2" grid
pattern

Follow-Up Questions

1. What were your greatest challenges in completing this project?

_____
_____
_____

2. Explain how this project will be used.

_____
_____
_____

3. Explain the procedure you used to transfer the design from the print to your pattern.

_____
_____
_____

Rubric for Make a Push Stick from a Pattern

Student Guidelines

Pattern layout

Possible points _____ Score _____

Did you follow the procedure in laying out grid and pattern?

_____

Materials and Resources

Possible points _____ Score _____

Did you supply a complete bill of materials, complete list of resources, and determine total cost of materials?

_____

Craftsmanship

Possible points _____ Score _____

Did you follow plans, sand out mill marks, round edges, and finish?

_____

Follow-up questions

Possible points _____ Score _____

Were all questions answered in a thoughtful manner?

_____

Total possible points _____ Score _____

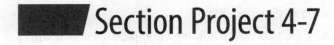

Section Project 4-7

Make a Routed Sign

There is an entire industry dedicated to making wood signs. Many communities have restrictions on lighted signs, creating an opportunity for cabinetmakers to create wood signs. Routers are typically used to make wood signs. This project will allow you to get familiar with using a router.

Resource Chapters

Chapter 10—Sketches, Mock-Ups, and Working Drawings
Chapter 12—Measuring, Marking, and Laying Out Materials
Chapter 26—Shaping
Chapter 29—Abrasives

Objectives

After completing this project, you should be able to:

- Properly select and change router bits.
- Demonstrate mastery of edge routing.
- Demonstrate mastery of freehand control of the router to follow a line.

Materials and Resources

- 3/4″ × 5″ × 16″ MDF or solid wood stock
- Router
- V-groove bit for routing words or numerals
- Roman ogee or 3/8″ round-over for adding a decorative edge

> **SAFETY NOTE**
>
> Before proceeding with this activity, you should have passed a general shop safety rule test, all specific safety tests on the tools used in this activity, and be certified by the instructor to use the tools and equipment needed for this activity.

Activity Procedure

1. Use graph paper to design a sign on 5″ × 16″ stock.

2. Lay out the full-size sign on white paper.

3. Glue the paper to the wood with spray adhesive. Don't use too much adhesive, or you will have trouble peeling off the paper.

4. Install the V-groove bit into the router and adjust to proper depth. Test this on scrap wood first.

5. Secure sufficient lighting and fasten material properly to the table. Put on hearing protection.

6. Turn on the router and carefully place it on your line.

7. Steadily follow the line and lift up the router when you reach the end.

   Tip: Allow the edges of your hand to rest on the material and control the router with your fingers. This will give you more control of the router. Make sure to angle your sight so you can see your line as you proceed with the cut.

8. When finished with the routing, change the router bit to the round-over or Roman ogee bit.

9. Adjust the bit depth to get the desired edge profile. Test this on scrap wood first.

10. Turn the router on and proceed counterclockwise around the sign. You will be forcing the router as far as it will go into the piece. The bearing on the end of the bit will keep you from cutting too deep.

11. When finished, check the edge to make sure you followed the wood edge all the way around. If you let the router come away, go back over that area.

> **Tip:** Depending on how much the router cuts, you may want to make a pass with the router at less-than-final depth to avoid splintering the wood and overloading the router. Some wood splinters more than others. Work carefully when finishing on the ends. Always make the final cut on an edge to minimize splintering out on the ends.

12. Peel the paper off and sand. If you used too much spray adhesive, you may have to sand the paper off.

13. Apply desired finish.

Final Project Requirements

- Shop sketch
- Plan of procedure
- Bill of materials
- Material cut sheet
- Final sign project
- Follow-up questions
- Student-produced activity print

Name _____

NAME :   DRAWING SCALE _____ = _____ SQUARES   DATE ___/___/___   FILE No. _____

Plan of Procedure

Minutes, Hours, Days, Weeks, or Months

Steps	1	2	3	4	5	6	7	8	9	10	11	12	13	14	15	16	17	18	19	20	21	22	23	24	25
1																									
2																									
3																									
4																									
5																									
6																									
7																									
8																									
9																									
10																									
11																									
12																									
13																									
14																									
15																									
16																									
17																									
18																									
19																									
20																									
21																									
22																									
23																									
24																									
25																									
26																									
27																									
28																									
29																									
30																									
31																									
32																									

Name _____

Bill of Materials

Project Name: _____

Quantity	Units	Material Description	Price/Unit	Extended Price
			Subtotal	
			Tax	
			Total	

Follow-Up Questions

1. What were your greatest challenges in completing this project?

   _____

   _____

   _____

   _____

2. Can you think of another method of marking the sign rather than gluing the paper on the board?

   _____

   _____

   _____

3. Did you finish your project on time? What processes went faster than you thought they would and what processes took longer? Did you use too much spray glue?

   _____

   _____

   _____

   _____

4. If you made $10.00 per hour, what would you be paid for this project?

   _____

   _____

5. Did you have any trouble keeping the router on line when you were routing your letters?

   _____

   _____

6. Explain how you kept the router steady when working freehand.

   _____

   _____

   _____

   _____

Rubric for Make a Routed Sign

Student Guidelines

Shop sketch
Possible points _____ Score _____

Assess your neatness, accuracy, dimensions, and correctness of views.

_____

Materials and resources
Possible points _____ Score _____

Did you supply a complete bill of materials, complete list of resources, and total cost of materials?

_____

Router
Possible points _____ Score _____

Did you change the bit without assistance? Is letter engraving steady, and edge shaping done properly?

_____

Craftsmanship
Possible points _____ Score _____

Were plans followed, mill marks sanded out, and corners broken properly?

_____

Follow-up questions
Possible points _____ Score _____

Were all questions answered in a thoughtful manner?

_____

Total possible points _____ Score _____

◢ Section Project 4-8

Make a Sanding Block

Properly finished workpieces usually involve some hand sanding. This step should be done carefully. Incorrect sanding in the final stages of project completion can make your work look substandard. There are many commercial sanding blocks on the market, but most cabinetmakers make their own. In this activity, you will learn how to tear sandpaper and make a sanding block.

Resource Chapter

Chapter 29—Abrasives

Objectives

After completing this project, you should be able to:

- Explain several methods for tearing sandpaper.
- Explain why using three layers of sandpaper is the best technique.
- Explain the abrading process.

Materials and Resources

- 3/4″ × 2 1/4″ × 5 1/2″ MDF or any suitable scrap wood that can be sawn
- One sheet of sandpaper, any grit

> **SAFETY NOTE**
>
> Before proceeding with this activity, you should have passed a general shop safety rule test, all specific safety tests on the tools used in this activity, and be certified by the instructor to use the tools and equipment needed for this activity.

Activity Procedure

1. Cut the MDF into 3/4″ × 2 1/4″ × 5 1/2″ dimensions.

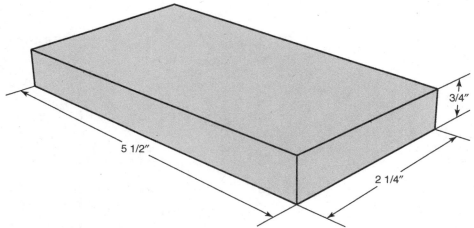

3/4″

5 1/2″

2 1/4″

Goodheart-Willcox Publisher

2. Fold a piece of 80 or 100 grit paper as shown in the following diagram. Make the fold so that the abrasive is on the outside of the fold and the plain paper is inside the fold.

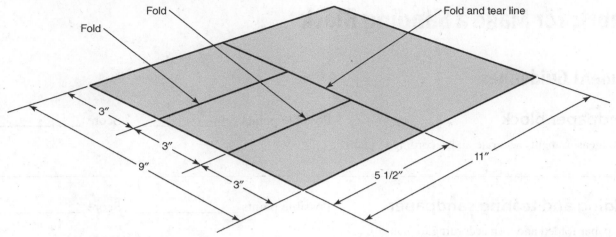

Goodheart-Willcox Publisher

3. Crease the fold sharply with your fingernail or a dowel. The paper will almost tear itself at this point. If you have trouble tearing the paper, make a small tear in the fold to start the tear. Then place the fold over the edge of a table and finish the tear. Now you have half a sheet.

4. Divide the half sheet into thirds and fold as the diagram above indicates. This creates the perfect size of paper for the sanding block you made in Step 1.

   Note: Another benefit to folding the sandpaper into thirds is that it can become a gripping tool. Place the folded paper between the workbench and a workpiece you are clamping to the bench. It will make the workpiece nearly immovable. Sandpaper folded into thirds for hand sanding is more stable in your hands. Compare with a piece of sandpaper that is less than one-half sheet. When you do not fold at all, you, your block, or hand will slide on the paper. When you double the sheet, it will slide between the two halves. But when you fold in thirds, it grips on both sides and the sheets will not slide between the folds. Experiment with this explanation and report your findings in the follow-up questions.

Final Project Requirements

- Sanding block
- Properly folded sandpaper
- Follow-up questions

Follow-Up Questions

1. What were your greatest challenges in completing this activity?

   _____

   _____

2. Explain how this project will be useful to you when building cabinets.

   _____

   _____

   _____

3. Did you have more control of the sandpaper when it was folded into thirds? Explain.

   _____

   _____

   _____

Rubric for Make a Sanding Block

Student Guidelines

Sandpaper block
Possible points _____ Score _____

Were measurements accurate and according to plans?

_____

Folding and tearing sandpaper
Possible points _____ Score _____

Was paper folded and torn according to plans?

_____

Follow-up questions
Possible points _____ Score _____

Were all questions answered in a thoughtful manner?

_____

Total possible points _____ Score _____

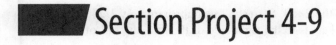 Section Project 4-9

Build a Sawhorse

In this project, you will build a sawhorse out of 2 × 4 lumber, using a portable circular saw and wide-blade handsaw. The challenge of this project is to keep the angles from becoming confusing. Work systematically to avoid that issue. Using a compound miter saw also helps simplify the process. The resulting sawhorse is strong, if the craftsmanship is good. A gusset or cross support can be added to make it even stronger.

Resource Chapters

Chapter 22—Sawing with Hand and Portable Power Tools
Chapter 23—Sawing with Stationary Power Machines

Objectives

After completing this project, you should be able to:

- Cut dimension lumber to length using a standard wide-blade crosscut handsaw. Pieces will be cut the correct length and perfectly square (90°) to the edges.
- Use a portable circular saw to cut a compound miter on dimension lumber. Cuts will be the specified angles when finished.
- Use a wide-blade handsaw or backsaw to make specialty cuts in dimension lumber.
- Properly set up the workstation for portable power equipment use.
- Determine how to clamp and support material being cut with portable power equipment and hand tools.

Materials and Resources

- Two 2″ × 4″ × 96″ studs (actual size 1 1/2″ × 3 1/2″ × 96″)
- Eleven 2″ wood screws
- Variable speed drill
- 3/16″ drill bit
- 3/8″ inch drill bit
- Phillips bit
- Wide-blade handsaw
- Portable circular saw
- Sliding T-bevel
- Square that can be used to lay out angles
- Pencil
- C-clamps

> **SAFETY NOTE**
>
> Before proceeding with this activity, you should have passed a general shop safety rule test, all specific safety tests on the tools used in this activity, and be certified by the instructor to use the tools and equipment needed for this activity.

Activity Procedure

1. Prepare workstation by running extension cords, gathering material, and planning how to support the material.

2. Set the angle on the portable circular saw to 15°.

3. Using the angle square, lay out a 10° angle on the end of the 2 × 4. Start the line approximately 1/2″ from the end to allow for some drop-off material. Be sure the saw kerf is to the right side of the line, the fall-off side of the material.

   Note: It is usually easier to obtain a true cut with a portable circular saw if you are not cutting right on the end of material.

4. Make a cut on the 10° layout line with the saw set to 15°. This will give you a compound miter cut. This is a cut that yields two angles on a piece of stock.

5. Measure 22 27/32″ from the longest point of the previous cut. Mark and lay out another 10° angle that is parallel to the first cut and make the cut. This time, the saw kerf should be on the left side of the layout line. Make the cut. The leg piece you are cutting is actually the fall-off material in this case.

6. If your last cut went smoothly, you should have the angle for the next leg already cut on the end of the remaining stock. If it did not cut smoothly, then simply move in about 1/2″ and lay out a 10° line as you did at first and make another cut. Then repeat Step 5. This will yield the second leg piece. You will need four of these pieces to form the four legs.

7. On the remaining stock, mark a 90° line with your square across the end of the piece, approximately 1/2″ from the end and cut this piece.

8. Measure 36″ from the end you just cut, mark another 90° line with your square, and make the cut.

   Note: For the next cut, you will want to rotate the 36″ piece so the cut is still on your right side. Otherwise, you would have to move around to the other side of the piece to make the cut, in order to keep the largest part of saw base plate supported.

9. Repeat Steps 3 to 8 on the second 2″ × 4″ × 96″ piece of material. You now have four legs cut and the two horizontal 36″ pieces for the sawhorse.

10. What remains is to cut the cheek cuts on the legs. Orient the legs in the position they will be in when assembled and mark the ends and side that will mate up to the horizontal piece. It is easy to confuse yourself on this and cut the wrong end, so lay it out on the floor to get the cuts straight in your mind.

11. Draw layout lines for the cheek cuts. Refer to the diagram.

12. Using a wide-blade handsaw, make the cheek cuts. There are various ways to support your work for this process. Be sure, however, to set up the material so that it will not move while cutting and so that you can get a full, comfortable stroke on the saw.

13. Drill three evenly spaced 3/16″ holes in the face of one of the 36″ 2 × 4s. One of the holes should be in the center of the length and two holes 6″ from the ends. All the holes should center in the width of the 2 × 4.

14. Counterbore the previous holes one inch deep with the 3/8″ bit.

15. Attach the two 36″ pieces together with 2″ wood screws.

16. Drill two 3/16″ pilot holes about 2″ from the top of each leg piece and 3/4″ from the edge. This will keep the wood from splitting when the screw is driven. Cut the legs.

17. Use the variable speed drill to drive the screws to attach the legs to the horizontal piece. Refer again to the diagram.

Final Project Requirements

- Completed sawhorse
- Follow-up questions

Name _____

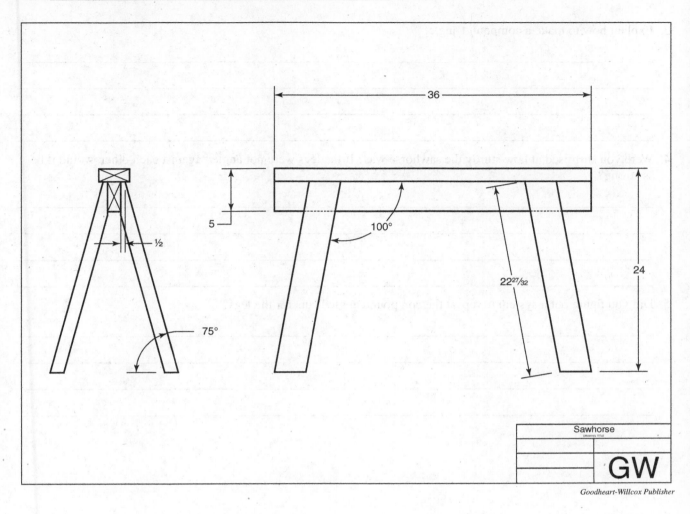

Sawhorse
DRAWING TITLE

GW

Goodheart-Willcox Publisher

Follow-Up Questions

1. What were your greatest challenges in completing this activity?

_____
_____
_____
_____

2. What is a saw kerf?

_____
_____
_____
_____

3. Explain how to make a compound angle.

_____

_____

_____

_____

4. Were you surprised at how sturdy the sawhorse was? If the legs were not angled against each other, would it be as strong?

_____

_____

_____

_____

5. Did you figure out a system to repeat the compound angled cuts for the legs?

_____

_____

_____

_____

Rubric for Build a Sawhorse

Student Guidelines

Measurements
Possible points _____ Score _____

Were measurements accurate according to plans?

_____

Assembly
Possible points _____ Score _____

Is the sawhorse accurately assembled?

_____

Cuts
Possible points _____ Score _____

Were cuts clean, accurate, and at proper angles?

_____

Follow-up questions
Possible points _____ Score _____

Were all questions answered in a thoughtful manner?

_____

Total possible points _____ Score _____

 Section Project 4-10

Sharpening a Chisel

Cabinetmakers must know how to maintain a keen edge on a chisel because dull chisels are useless and dangerous. In this activity, you will practice sharpening a chisel. A properly sharpened chisel is razor sharp.

Resource Chapters

Chapter 24—Surfacing with Hand and Portable Power Tools
Chapter 39—Sharpening

Objectives

After completing this project, you should be able to:

- Explain the concept of hollow ground.
- Determine when a chisel needs only honing or needs both grinding and honing.
- True the end of a chisel.
- Properly grind a chisel.
- Properly hone a chisel.
- Explain the safety principles to follow when sharpening and working with chisels.

Materials and Resources

- Grinder with medium and fine stone and adjustable tool supports
- Dull chisel
- Try square
- Cup of cooling water
- Sharpening stones, 800 to 6000 grit

> **SAFETY NOTE**
>
> Before proceeding with this activity, you should have passed a general shop safety rule test, all specific safety tests on the tools used in this activity, and be certified by the instructor to use the tools and equipment needed for this activity.

Activity Procedure

1. Make a profile drawing of how you will sharpen the chisel. Make your drawing similar to **Figure 24-5** in the text. The drawing is a plane iron, not a chisel, but the profile is the same.

2. Write a plan of procedure.

3. Determine if chisel needs grinding and honing or honing only.

4. Check the end of the chisel with a square to ensure the end is perfectly square.

5. Use the sharpening stone with the coarsest grit to true the end. Hold the chisel at a 90° angle to the stone and hone in a figure eight pattern.

6. If the edge is far from square, you may have to grind the end square. If you have to do this, quite a bit of grinding will be required to get the hollow ground edge back.

7. Once the end is square, set the tool rest so that the grinding wheel will grind a 25° angle on the end of the chisel.

8. For safety, the tool rest should not be more than 1/8″ inch away from the grinding wheel.

9. Move the chisel carefully back and forth, making sure to keep the grinding even. Watch the tip carefully. Do not allow it to overheat. Dip the edge frequently into the water to keep the edge cool.

   Note: A slow grinder will reduce overheating. Slow grinders are rated at 1800 rpm or less. When a tool is overheated during grinding, it will lose the tempering properties in the metal thus making it dull faster.

10. Grind the chisel until the ground surface meets the tip and you can no longer see a shine on the tip. Look directly at the end with light behind you.

11. Now you are ready to start the honing process. You must keep the chisel at a 30° angle while honing. Use a honing fixture, if available. Otherwise, get a feel for the angle and maintain that angle while honing. Honing by feel takes practice. You must maintain the same angle or you will not get satisfactory results.

12. Test the chisel sharpness by following some of the suggested tests discussed in the textbook.

Final Project Requirements

- Profile drawing
- Plan of procedure
- Sharp chisel
- Follow-up questions

Follow-Up Questions

1. What were your greatest challenges in completing this activity?

   _____

   _____

   _____

2. Why is it important to have a sharp chisel?

   _____

   _____

   _____

3. Did you have all the equipment needed to perform the task?

   _____

   _____

   _____

Rubric for Sharpening a Chisel

Student Guidelines

Profile drawing

Possible points _____ Score _____

Assess neatness, accuracy, dimensions, and correctness of views.

_____

Plan of procedure

Possible points _____ Score _____

Was there a detailed list of steps?

_____

Craftsmanship

Possible points _____ Score _____

Did you use proper grinding and honing techniques?

_____

Follow-up questions

Possible points _____ Score _____

Were all questions answered in a thoughtful manner?

_____

Total possible points _____ Score _____

 **Section Project 4-11**

Make a Sign with a CNC Router

A CNC router can be used to do just about any operation that can be done with traditional hand tools and portable power and stationary power machines. The CNC router has revolutionized the making of wood signs. What used to take hours can now be done in minutes. Have fun designing and making your sign.

Resource Chapters

Chapter 6—Components of Design
Chapter 11—Creating Working Drawings
Chapter 28—Computer Numerically Controlled (CNC) Machinery

Objectives

After completing this project, you should be able to:
- Make a simple rectangular shape on a computer using CAD software.
- Export a DXF file from CAD software to CAD/CAM software.
- Generate the tool paths and machine code file in the CAD/CAM software.
- Set up and run the CNC machine to make a simple sign.

Materials and Resources

- CNC router
- CAD software
- 3/4″ MDF blank size of your design

> **SAFETY NOTE**
> Before proceeding with this activity, you should have passed a general shop safety rule test, all specific safety tests on the tools used in this activity, and be certified by the instructor to use the tools and equipment needed for this activity.

Activity Procedure

1. Sketch out ideas on a piece of graph or plain paper.

2. Draw the design using CAD software.

3. Make a DXF file of the drawing.

4. Import the file into the CAD/CAM software and design the tool paths.

5. Export the file to the machine controller.

6. Fixture the wood blank onto the bed of the router.

7. Zero the router.

8. Start the program and cut the sign.

Final Project Requirements

- Preliminary scale sketches on paper
- CAD drawing of your design
- Machine code file
- Final product made on the CNC router
- Follow-up questions

Follow-Up Questions

1. What were your greatest challenges in completing this project?

   _____

   _____

   _____

   _____

2. Explain how the finished product will be used.

   _____

   _____

   _____

   _____

3. How many hours did you spend on this project?

   _____

4. Why does the machine have to be zeroed in?

   _____

   _____

   _____

   _____

Rubric for Make a Sign with a CNC Router Project

Student Guidelines

Sketch
Possible points _____ Score _____

Did you create a detailed, accurate description and graphic representation of the final sign?

_____

CAD drawing
Possible points _____ Score _____

Did you successfully complete the drawing? Was time on computer spent efficiently? Did time spent demonstrate understanding?

_____

Tool paths
Possible points _____ Score _____

Were tool paths efficient and practical?

_____

Program execution
Possible points _____ Score _____

Was part mounted properly in the machine? Was machine zeroed in properly the first time?

_____

Follow-up questions
Possible points _____ Score _____

Were all questions answered in a thoughtful manner?

_____

Total possible points _____ Score _____

Section Project 4-12

Veneering

The idea behind veneering is to make a less expensive wood look and feel like a more expensive wood. In this activity, you will practice veneering techniques on a 6″ × 6″ × 3/4″ MDF core.

Resource Chapters

Chapter 22—Sawing with Hand and Portable Power Tools
Chapter 23—Sawing with Stationary Power Machines
Chapter 30—Using Abrasives and Sanding Machines
Chapter 31—Adhesives
Chapter 32—Gluing and Clamping
Chapter 34—Overlaying and Inlaying Veneer

Objectives

After completing this project, you should be able to:

- Understand the veneering process.
- Cut veneer pieces and join them with veneer tape.

Materials and Resources

- Veneer press or vacuum bag press
- 3/4″ × 6″ × 6″ MDF
- 6″ × 6″ backer board
- Veneer pieces of different kinds of wood
- Veneer tape or high-quality masking tape
- Edgebanding
- Edgebanding machine or simple veneer iron or household electric iron
- Veneer saw
- Hand plane
- Veneer glue
- Wax paper

SAFETY NOTE

Before proceeding with this activity, you should have passed a general shop safety rule test, all specific safety tests on the tools used in this activity, and be certified by the instructor to use the tools and equipment.

Activity Procedure

1. Cut veneer with a veneer saw or utility knife to a rough size of 3 1/8″ × 3 1/8″.

2. Make a trimming clamp as shown in **Figure 34-7A** of the text.

3. Plane each edge to make the pieces exactly 3″ × 3″. Make sure the pieces are perfectly square. Glue sandpaper on the surfaces of your trimming clamp to ensure the veneer does not slip. Make a 3″ × 3″ block to use to mark the finished size of the veneer.

4. Lay out the veneer on the 6″ × 6″ MDF in one of the patterns shown in **Figure 34-5** of the text. Make sure the veneer joints are perfect with no gaps.

5. Tape the joints with veneer tape as shown in **Figure 34-8** of the text.

6. Set up the veneer press or the vacuum bag by getting the cauls ready. Wax paper should separate the cauls from the pieces to be clamped. Otherwise, you might glue your project to the cauls.

7. Apply glue to the backer board and attach to the 6″ × 6″ piece, and then apply the glue to the surface of the MDF and attach the veneer assembly.

8. Set the "sandwich" into the press or vacuum bag between the wax paper and the cauls.

9. Apply pressure or the vacuum.

10. Let set in the press at least two hours.

11. Let the assembly dry 24 hours before working.

12. Clean glue with scrapers, and then sand. Be careful, if scraping, to not damage veneer surface. If there is a lot of excess glue, note that you have used too much glue and can use less next time.

13. Carefully prepare the edges for the edge trim. This can be done on a edge sander or with a hand plane. A setup similar to the one shown in **Figure 30-6** of the text can be used to sand the edges; be careful because you don't want to round the edges.

14. Use veneer edge trim with hot-melt glue coating to edge trim the four edges of the 6″ × 6″ piece. One option is to make 1/16″ edge trim and glue with contact adhesive.

15. Sand the edges to bring the edge trim flush with the surfaces.

Final Project Requirements

- 6″ × 6″ piece
- Follow-up questions

Follow-Up Questions

1. What were your greatest challenges in completing this activity?

_____

_____

_____

2. Explain how the skills developed in the activity can be used in the design and construction of cabinets.

_____

_____

_____

3. What veneer placement pattern did you use to order your four veneer pieces on the MDF? See **Figure 34-5** in the text. Explain your choice.

_____

_____

_____

Rubric for Veneering

Student Guidelines

Veneering
Possible points _____ Score _____

Are veneer joints tight?

_____

Veneer pattern
Possible points _____ Score _____

Did you lay out squares in an attractive manner?

_____

Edgebanding
Possible points _____ Score _____

Is edgebanding solidly glued to the edges, and are corners trimmed correctly?

_____

Sanding and finish
Possible points _____ Score _____

Are imperfections puttied using good technique? Are edges and corners sanded well?

_____

Follow-up questions
Possible points _____ Score _____

Were all questions answered in a thoughtful manner?

_____

Total possible points _____ Score _____

416   Modern Cabinetmaking Lab Workbook

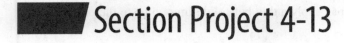 ## Section Project 4-13

Edge Trim

Manmade materials provide many advantages to production cabinetmaking. A disadvantage, however, is that the edges of the material are unattractive in most cases. Therefore, it is necessary to "dress up" the edges to hide the unsightly edges of plywood, MDF, and melamine.

Resource Chapters

- Chapter 16—Manufactured Panel Products
- Chapter 22—Sawing with Hand and Portable Power Tools
- Chapter 23—Sawing with Stationary Power Machines

Objectives

After completing this project, you should be able to:

- Apply an edgeband that has hot-melt glue coating.
- Be able to apply a veneer edgeband that is not pre-glued.
- Use shop equipment to cut material to specified sizes.

Materials and Resources

- 4″ × 5″ maple plywood
- 1/4″ × 13/16″ × 4″ solid wood maple edgeband
- Maple edgeband with hot-melt glue coating
- Veneer wood edgeband (yellow glue applied and clamped)
- 3/4″ × 1 1/4″ × 4″ solid maple
- Small quick clamps and glue cauls
- Veneer iron or regular household iron
- Wax paper

SAFETY NOTE

Before proceeding with this activity, you should have passed a general shop safety rule test, all specific safety tests on the tools used in this activity, and be certified by the instructor to use the tools and equipment.

Activity Procedure

1. Cut the 4″ × 5″ plywood.

2. Make the 1/4″ × 13/16″ × 4″ solid wood edgeband.

3. Make the 3/4″ × 1 1/4″ solid wood edge trim.

4. Cut a 5 1/8″ piece of the edge trim with the hot-melt glue coating off the roll.

5. Cut a 5 1/8″ piece of the veneer edgeband off the roll.

6. Attach the edge trim with the hot-melt glue coating to a 5″ edge. Make sure to center the banding so that a little sticks past both surfaces. Heat with an iron and press on it with a J-roller or a wooden press-down stick until it cools.

7. The edges that stick past the surfaces must be edge trimmed with a router with a flush trim bit installed, and carefully sanded. If care is taken, you will hide the edge of the plywood so well that it will be hard to tell it is not solid wood.

8. The next step requires glue, clamps, and wax paper strips. Apply a thin but solid film of glue to the other 5″ edge of the plywood and veneer edgeband.

9. Press the two parts together. The quick setting yellow glue will begin to set the edgeband and hold it in place. Place wax paper strips on the edgeband and then the caul. Clamp the assembly with two clamps. After about an hour, you can remove the clamps. It will be 2–3 hours before you can actually work on the piece again. The edges will be edge trimmed with a router and sanded as was done on the edge with the hot-melt glue coating.

10. Next, cut the groove in the edge of the plywood as the plans indicate.

11. Cut the 1 1/4″ × 4″ piece out of solid maple stock.

12. Cut the tongue on the 1 1/4″ × 4″ piece as plans indicate. (This step would be more easily done in a larger piece and then cut to length and distributed). You may want to work with your other classmates to accomplish this.

13. Cut the 1/4″ × 13/16″ piece out of solid maple stock.

14. Apply glue to the 1/4″ × 4″ piece and the 1 1/4″ × 4″ piece and the edges of the plywood. You are going to clamp this together at the same time. Place wax paper between the caul and the 1/4″ piece and clamp with the two clamps. You will not need a caul for the 1 1/4″ piece.

15. Let the assembly dry a minimum of two hours and sand the edges flush with the face. The 1/4″ piece may need to be edge trimmed and then sanded.

Final Project Requirements

- The piece completed as the plans indicate
- Follow-up questions

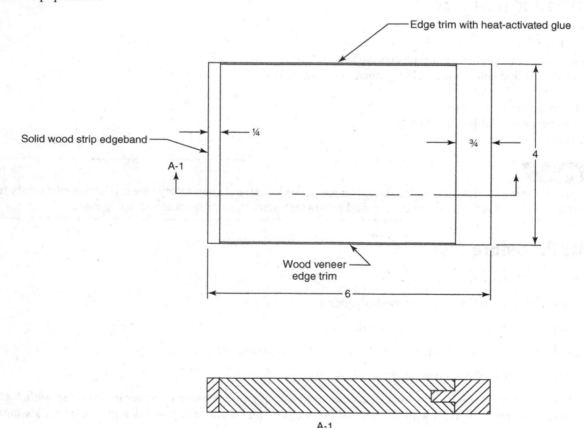

Goodheart-Willcox Publisher

Follow-Up Questions

1. What were your greatest challenges in completing this project?

   _____

   _____

   _____

2. Explain which edgebanding technique you liked most, and explain why.

   _____

   _____

   _____

Rubric for Edge Trim

Student Guidelines

All wood edge trim

Possible points _____ Score _____

Are the tongue and groove tight, sanded flush with surface, with no glue showing?

_____

1/4″ edge trim

Possible points _____ Score _____

Did you ensure that the trim is sanded flush with face, no glue is showing, and there is solid mating of band to plywood?

_____

Veneer edge trim

Possible points _____ Score _____

Did you ensure there are no hollow spots, corners are sanded well, and no glue is showing?

_____

Edge trim with hot-melt glue coating

Possible points _____ Score _____

Did you ensure there are no hollow spots, corners are sanded well, and no glue is showing?

_____

Craftsmanship

Possible points _____ Score _____

Did you follow plans, sand out mill marks, and ensure corners are broken properly?

_____

Follow-up questions

Possible points _____ Score _____

Were all questions answered in a thoughtful manner?

_____

Total possible points _____ Score _____

 Section Project 4-14

Practice Applying Plastic Laminate

In this project, you will follow the plans to make a simulated kitchen countertop out of particleboard. Plastic laminate will be applied to the surface and front edge as per plans.

Resource Chapters
- Chapter 17—Veneers and Plastic Overlays
- Chapter 33—Bending and Laminating

Objectives
After completing this project, you should be able to:
- Apply plastic laminate to panel using contact cement.
- Apply plastic laminate to an edge.
- Apply heat-activated laminate.

Materials and Resources
- 4″ × 6″ industrial particleboard
- 1 1/2″ × 6″ industrial particleboard
- 4 1/4″ × 6 1/2″ plastic laminate
- 1 3/4″ × 6″ plastic laminate
- Contact cement (water-based cement if ventilation is suspect)
- Disposable 1″ brushes
- Nail gun
- J-roller
- Yellow glue

> **SAFETY NOTE**
>
> Before proceeding with this activity, you should have passed a general shop safety rule test, all specific safety tests on the tools used in this activity, and be certified by the instructor to use the tools and equipment.

Activity Procedure

1. Cut pieces as per plans.

2. Apply yellow wood glue to one long edge of the 4″ × 6″ industrial particle board and nail the 1 1/2″ × 6″ to the 4″ × 6″ as per plans.

3. Sand joint flush with surface.

4. Clean all dust.

5. Apply cement to 1 1/2″ edge and the 1 3/4″ × 6″ plastic laminate and let dry until you can touch it and it does not feel wet or stick to your skin. This is usually 10–15 minutes, depending on temperature, humidity, and brand of cement.

6. Carefully apply the 1 1/2″ × 6″ piece to the edge. Once the glued surfaces touch, it is stuck and cannot be moved.

7. Use a router with a flush trim bit to trim the laminate flush with all edges.

8. Use a sander to flush the laminate and the edge flush with the surface of the 4″ × 6″ piece. Anything out of square here will show through the surface laminate when applied.

9. Clean all dust from the surface.

10. Apply cement to both the surface 4″ × 6″ particleboard and the contact surface of the 4 1/2″ × 6 1/2″ laminate.

11. Let dry as before.

12. Carefully apply as before and press down with a J-roller.

13. Flush trim all edges with a flush trim bit.

14. Smooth the corner formed at the joint of the two laminates with 600 grit wet or dry silicon carbide sandpaper. This makes for a more comfortable finished corner.

Final Project Requirements
- Laminated project
- Follow-up questions

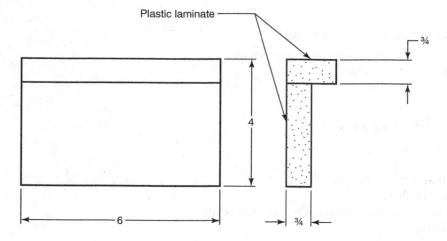

Goodheart-Willcox Publisher

Follow-Up Questions

1. What were your greatest challenges in completing this project?

_____

_____

_____

2. Explain how plastic laminate is made.

_____

_____

_____

3. Explain how contact cement works.

_____

_____

_____

4. What kind of contact cement did you use?

_____

Name _____

Rubric for Applying Plastic Laminate

Student Guidelines

Construction

Possible points _____ Score _____

Did you ensure neatness, accuracy, dimensions, and correctness of views?

_____

Lamination

Possible points _____ Score _____

Did you complete a bill of materials, complete a list of resources, and figure total cost of materials?

_____

Laminate finishing

Possible points _____ Score _____

Did you keep a detailed list of steps to build the project? Did you keep accurate time records of estimated time and actual time for each step?

_____

Follow-up questions

Possible points _____ Score _____

Were all questions answered in a thoughtful manner?

_____

Total possible points _____ Score _____

Notes

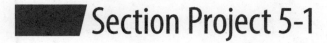 Section Project 5-1

Surfacing Stock for Cabinet Face Frames

The two uses of the planer are preparing rails and stiles for the face frame of cabinets and preparing solid stock for raised panel door construction. In this activity, you will plane face frame material for a project to a precise thickness and width and evaluate the performance of the planer.

Resource Chapters

Chapter 25—Surfacing with Stationary Machines
Chapter 41—Frame and Panel Construction

Objectives

After completing this project, you should be able to:

- Use the planer to prepare stock for use in face frame construction.
- Diagnose poor performance of a planer or jointer.
- Explain why it is most efficient to prepare rail and stile stock for several cabinets at one time.

Materials and Resources

- Solid stock, enough for face frame material for several cabinets
- Planer
- Table saw
- Calipers (preferably digital)

> **SAFETY NOTE**
>
> Before proceeding with this activity, you should have passed a general shop safety rule test, all specific safety tests on the tools used in this activity, and be certified by the instructor to use the tools and equipment.

Activity Procedure

1. Plane stock to a precise thickness needed for your rails and stiles, usually 3/4″.

2. If one edge of the stock is not straight line ripped, it must be made straight on a jointer or by some other safe technique.

3. Rip the stock to 1/8″ over the standard width that is used in your shop for rails and stiles.

4. Set the planer to cut the width of the material to within 1/16″ of the final width.

5. Turn the stock on edge and hold about 4–6 pieces together and run through the planer.

6. Repeat this process until all of your material is surfaced on one edge. This process removes saw marks.

7. Inspect the material to make sure the planer is adjusted properly. If not, make the proper adjustments as the text and planer owner's manual instruct.

8. Turn all the material over and repeat this process on the other edge, bringing the material to the exact width measurement.

Final Project Requirements

- Style and rail material finished to thickness and width, ready for use in face frame cabinet construction
- Follow-up questions

Follow-Up Questions

1. What were your greatest challenges in completing this activity?

   _____

   _____

   _____

2. Why is the material for stiles and rails prepared in bulk?

   _____

   _____

   _____

3. How many total working hours did it take to prepare the stock from start to finish?

   Number of workers _____ × _____ hours = _____ working hours.

4. Did the planer perform perfectly? Explain your answer.

   _____

   _____

   _____

Name _____

Rubric for Surfacing Stock for Cabinet Face Frames

Student Guidelines

Jointer work

Are the edges straight and square to face?

Possible points _____ Score _____

_____

Efficiency

What were the total working hours?

Possible points _____ Score _____

_____

Planer work

Did you properly evaluate planer performance and follow procedures?

Possible points _____ Score _____

_____

Follow-up questions

Were all questions answered in a thoughtful manner?

Possible points _____ Score _____

_____

Total possible points _____ Score _____

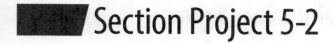

Section Project 5-2

Trinket Box

This is a beginner project in cabinetmaking. This project is a good practice for the beginning cabinetmaker because there are similarities to building a drawer.

Resource Chapter

Chapter 40—Case Construction

Objectives

After completing this project, you should be able to:

- Build a cabinetmaking project.
- Organize the work of cabinetmaking.
- Use a plan of procedure to keep the work on track.
- Use the resources in a typical cabinet shop to complete a cabinetmaking project.

Materials and Resources

- 2 1/2 board feet of solid wood
- 1/8" 7" × 12" tempered hardboard
- 3/4" × 3/4" brass butt hinge
- Box lid latch
- Sandpaper
- Stain
- Clear finish

SAFETY NOTE

Before proceeding with this activity, you should have passed a general shop safety rule test, all specific safety tests on the tools used in this activity, and be certified by the instructor to use the tools and equipment.

Activity Procedure

1. Study the prints and plan of procedure.

2. A study of Chapter 40 will give you some ideas of how to proceed with the construction of this project, especially Sections 40.3.6 to 40.4.1 and **Figures 40-9** and **40-10**.

3. Use the bill of materials to gather your materials.

4. Follow the procedure on the plan of procedure sheet provided.

Final Project Requirements

- Completed trinket box
- Plan of procedure with updated time blocked in pencil
- Material cut sheet with checks as process was followed
- Follow-up questions

Bill of Materials

Project Name: Trinket Box

Quantity	Units	Material Description	Price/Unit	Extended Price
2 1/2′	BF	Solid wood, builder's choice		
1	each	1/8″ thick, 7″ x 12″ tempered hardboard		
1	pair	3/4″ x 3/4″ brass butt hinge		
1	each	Box lid latch		
1	sheet	Sandpaper		
1	quart	Stain		
1	quart	Clear finish		
			Subtotal	
			Tax	
			Total	

Trinket Box Plan of Procedure

Steps	1	2	3	4	5	6	7	8	9	10	11	12	13	14	15	16	17	18	19	20	21	22	23	24	25
1 Study plans	█																								
2 Fill out cut list	█																								
3 Rough cut all materials		█																							
Sides																									
4 Square a face and plane to thickness		█																							
5 Square an edge and rip to width		█																							
6 Cut drawer groove			█																						
7 Cut miters on ends of side pieces			█																						
8 Measure inside length of the groove on long and short sides			█																						
9 For bottom, cut to above measurements minus 1/16"			█																						
10 Cut the feet shape on the bottom				█																					
11 Apply glue & assemble sides together with bottom				█																					
Box Assembly																									
12 Let dry overnight				█																					
13 Square the top edge					█																				
Top																									
14 Cut the top to 1/4" of size of the box				█																					
15 Glue and clamp to the box assembly					█																				
Box Assembly																									
16 Let dry overnight					█																				
17 Flush trim or hand plane top flush with sides						█																			
18 Sand with 120 grit garnet paper						█																			
19 Cut top off with table saw						█																			
Box and Lid																									
20 Mark lid and box so they stay oriented						█																			
21 Square mating edges removing saw marks							█																		
22 Lay out and mortise the hinge gain for butt hinges							█																		
23 Install hinges and door latch							█																		
24 Adjust until lid mates well with box								█																	
25 Take hinges off for finishing process								█																	
26 Remove any dents								█																	
27 Putty cracks and holes								█																	
28 Sand with 180 grit garnet paper								█																	
29 Apply finish									█																
30 Reinstall hinges and box latch																									
31 Gray is the suggested scheduling. Keep a record of the actual time scheduling by marking with pencil.																									

Hours

Name _____

Material Cut Sheet

Part Name	Material	Number Pieces	Rough Measurements			Cut Process Completed	Finish Measurements			Cut Process Completed
			Thickness	Width	Length		Thickness	Width	Length	

Trinket Box

Note: All material is 1/2 inch thick
Builder choice on latch hardware

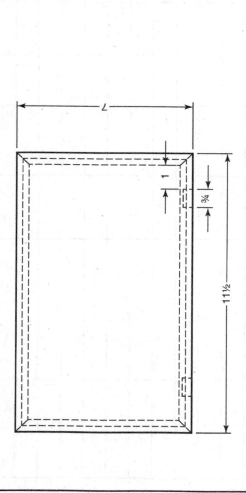

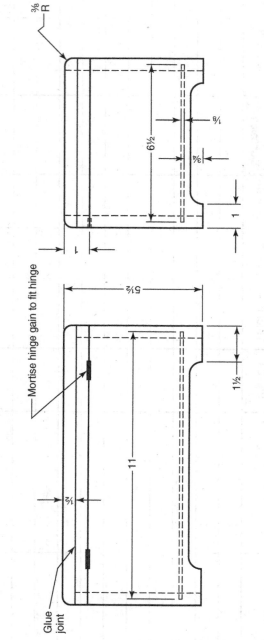

Mortise hinge gain to fit hinge

Glue joint

Follow-Up Questions

1. What were your greatest challenges in completing this project?

   _____

   _____

2. Explain how you will use this project.

   _____

   _____

3. Did you finish your project on time? What processes went faster than you thought they would, and what processes took longer?

   _____

   _____

4. Did you stay within your budget (this is what you figured on the bill of materials form) on the materials? What was the total cost of the materials for this project?

   _____

   _____

5. If you made $10.00 per hour, what was the cost of the labor on this project?

   _____

6. Explain the process that was used to mortise the hinge gains.

   _____

   _____

Rubric for Trinket Box

Student Guidelines

Plan of procedure

Possible points _____ Score _____

Did you keep accurate records on the plan of procedure by shading in the square on the sheet for actual time?

_____

Cut list

Possible points _____ Score _____

Is the cut list form filled out correctly and completely?

_____

Craftsmanship

Possible points _____ Score _____

Did you follow plans, sand out mill marks, and ensure corners are broken properly? Are edges flush at joints?

_____

Follow-up questions

Possible points _____ Score _____

Were all questions answered in a thoughtful manner?

_____

Total possible points _____ Score _____

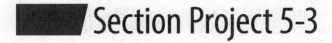

Section Project 5-3

Bedside Face Frame Cabinets

This project is designed to give you practice in case construction with a face frame method of constructing cabinets. It features basic carcase construction face frame techniques, recessed back, and simple panel door and drawer construction. The rail and stile joinery for the face frame is not detailed in the plans because of the many different joints that are used to fasten the rails to the stiles. This is the cabinetmaker's choice. These techniques should be discussed and the best one for your shop chosen. There are also several options for the joinery in the drawer construction.

Resource Chapters

Chapter 9—Production Decisions
Chapter 22—Sawing with Hand and Portable Power Tools
Chapter 23—Sawing with Stationary Power Machines
Chapter 37—Joinery
Chapter 40—Case Construction
Chapter 41—Frame and Panel Construction

Objectives

After completing this project, you should be able to:

- Master basic case construction skills.
- Master face frame construction techniques.
- Make a panel cabinet door.
- Use European concealed hinges.
- Plan and execute the construction of a basic cabinet.
- Demonstrate skill in drawer making.
- Use all the equipment and tools of the shop to the task of building cabinets.

Materials and Resources

- Solid 3/4″ stock for face frame
- Cabinet grade 3/4″ 4′ × 8′ MDF or plywood
- 1/8″ hardboard
- Drawer stock
- Drawer guides
- European concealed hinges
- 5 mm shelf pins
- Finish and sandpaper

> **SAFETY NOTE**
>
> Before proceeding with this activity, you should have passed a general shop safety rule test, all specific safety tests on the tools used in this activity, and be certified by the instructor to use the tools and equipment.

Activity Procedure

1. Study plans. Your instructor may want to make the stiles and rail material and cut the major parts of the cabinet as a class.

2. Make bill of materials and figure the estimated cost.

3. Make cut list.

4. The plans show the edges of the trim with a radius. Another option would be to add a shaped edge other than a radius. Discuss this with your instructor to find out what capabilities your shop has for shaping edges. Specify this in your plan of procedure.

5. Fill out the plan of procedure. Refer to Chapter 9, Sections 9.3 and 9.4.

6. Follow plan of procedure and keep time records on the plan of procedure form.

7. When making the drawer, make sure the drawer is exactly 1″ less than the distance between the stiles at the drawer opening. Most drawer glides require this; check manufacturers' installation instructions.

Final Project Requirements

Carefully study the project management forms for the Trinket Box project to see a model of how to do the wording and time chart for the following forms.

- List of resources needed, tools needed, and bill of materials
- Plan of procedure
- Material cut sheet
- 4′ × 8′ layout plan and cut list sheet
- Bill of materials
- Completed cabinet
- Follow-up questions

Name _____

4′ × 8′ Layout Plan and Cut List

Period _____ Date _____

Name or Job _____

Part Label	Pieces	Thickness	Width	Length	Square Inches	Square Feet

Bill of Materials

Project Name: _____

Quantity	Units	Material Description	Price/Unit	Extended Price
			Subtotal	
			Tax	
			Total	

Name _____

Plan of Procedure

Minutes, Hours, Days, Weeks, or Months

Steps	1	2	3	4	5	6	7	8	9	10	11	12	13	14	15	16	17	18	19	20	21	22	23	24	25
1																									
2																									
3																									
4																									
5																									
6																									
7																									
8																									
9																									
10																									
11																									
12																									
13																									
14																									
15																									
16																									
17																									
18																									
19																									
20																									
21																									
22																									
23																									
24																									
25																									
26																									
27																									
28																									
29																									
30																									
31																									
32																									

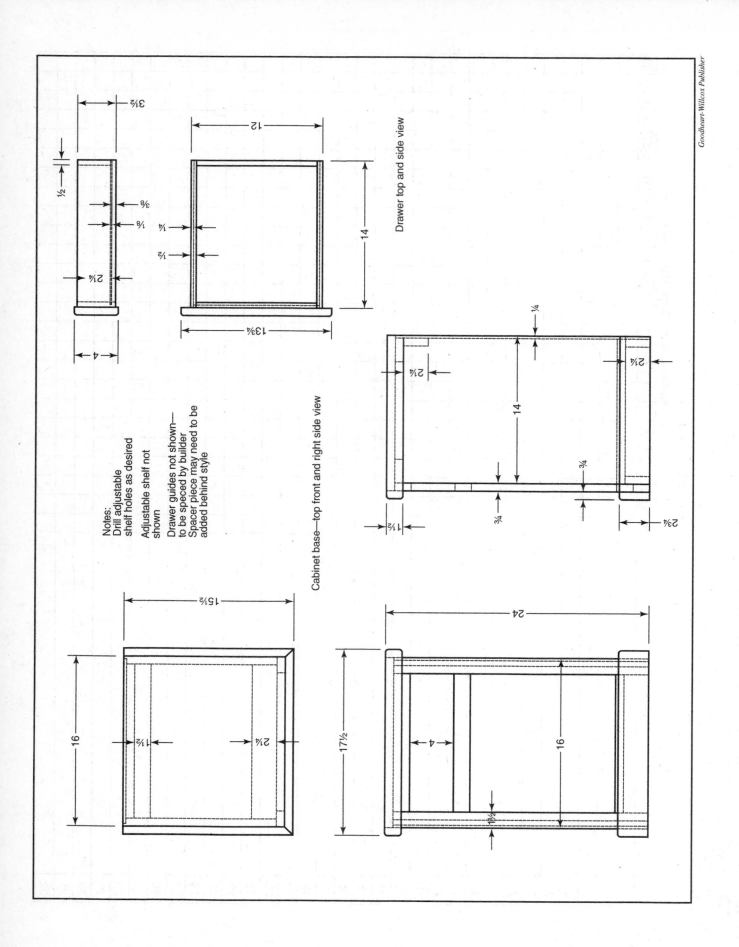

Drawer top and side view

Cabinet base—top front and right side view

Notes:
Drill adjustable
shelf holes as desired

Adjustable shelf not
shown

Drawer guides not shown—
to be speced by builder
Spacer piece may need to be
added behind style

Goodheart-Willcox Publisher

Follow-Up Questions

1. What were your greatest challenges in completing this project?

   _____

   _____

   _____

2. Explain how you will use this project.

   _____

   _____

   _____

3. Did you finish your project on time? What processes went faster than you thought they would, and what processes took longer?

   _____

   _____

   _____

4. Did you stay within your budget (this is what you figured on the bill of materials form) on the materials? What was the total cost of the materials for this project?

   _____

   _____

5. How many hours did it take you to complete this project?

   _____

   _____

6. If you made $10.00 per hour, what was the cost of the labor on this project?

   _____

7. What joints did you use to complete your face frame? Explain why.

   _____

   _____

Rubric for Bedside Face Frame Cabinets

Student Guidelines

Materials and resources

Possible points _____ Score _____

Did you complete a bill of materials, fill out the cut list correctly, and figure total cost of materials?

_____

Plan of procedure

Possible points _____ Score _____

Did you keep a detailed list of steps to build the project? Did you keep accurate time records of estimated time and actual time for each step?

_____

Craftsmanship

Possible points _____ Score _____

Did you follow plans, sand out mill marks, and ensure corners are broken properly? Are edges flush at joints? Are measurements accurate, and the cabinet square?

_____

Follow-up questions

Possible points _____ Score _____

Were all questions answered in a thoughtful manner?

_____

Total possible points _____ Score _____

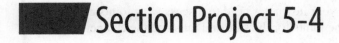 Section Project 5-4

Wall Cabinets—Face Frame and Carcase Construction

This project is designed to give you practice in the basic face frame method of constructing cabinets. It features basic frame techniques, recessed back, and simple carcase construction. The rail and stile joinery is not detailed in the plans because of the many different joints that can be selected to complete the face frame. A door is recommended but is not shown. The door may only cover the larger upper opening in the cabinet and leave the lower portion open. This is optional. There are so many ways to make a door; it is recommended that you study Chapter 43 and draw up your own design.

Resource Chapters

Chapter 9—Production Decisions
Chapter 22—Sawing with Hand and Portable Power Tools
Chapter 23—Sawing with Stationary Power Machines
Chapter 37—Joinery
Chapter 40—Case Construction
Chapter 41—Frame and Panel Construction
Chapter 43—Doors

Objectives

After completing this project, you should be able to:

- Master basic carcase construction skills.
- Master face frame construction techniques.
- Make cabinet doors.
- Use European concealed hinges.
- Plan and execute the construction of a basic cabinet.
- Use all the equipment and tools of the shop to the task of building cabinets.

Materials and Resources

- Cabinet grade 3/4″ 4′ × 8′ MDF or plywood
- Solid wood stock for face frame
- 1/4″ back material
- European concealed hinges and plates
- Finish supplies and sandpaper
- Shelf pins

SAFETY NOTE

Before proceeding with this activity, you should have passed a general shop safety rule test, all specific safety tests on the tools used in this activity, and be certified by the instructor to use the tools and equipment.

Activity Procedure

1. Study plans.

   A. Your instructor may want to make all the stiles and rail material and cut the major parts of the cabinet as a class.

   B. You will need to design your door before starting your plan of procedure.

2. Make bill of materials and figure the estimated cost.

3. Make cut list.

4. Before the plan of procedure can be completed, you and your instructor need to discuss what joints can be constructed in your shop. You will then need to choose the joints for the stile and rails and the cabinet sides.

5. Fill out plan of procedure. Refer to Chapter 9, Sections 9.3 and 9.4.

6. You may want to discuss adding shaped edges. Discuss with your instructor the capabilities of your shop to do this. Specify this in your plan of procedure.

7. Follow plan of procedure and keep time records.

Final Project Requirements
- Bill of materials
- 4′ × 8′ layout plan and cut list sheet (filled out)
- Material cut sheet
- Plan of procedure
- Shop drawings of door design
- Completed wall cabinet
- Follow-up questions

Name _____

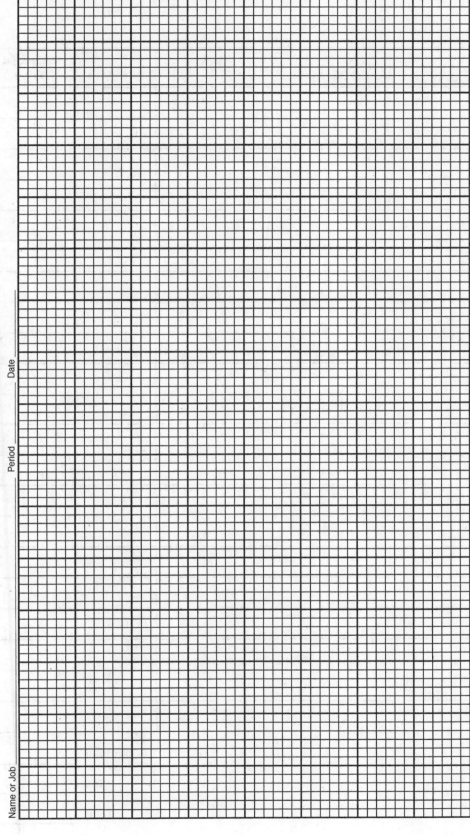

4' × 8' Layout Plan and Cut List

Name or Job _____ Period _____ Date _____

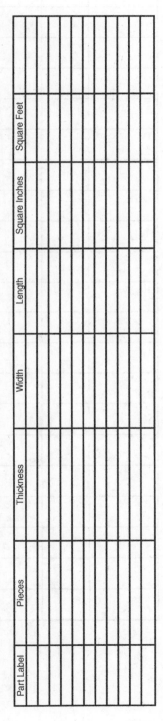

Part Label	Pieces	Thickness	Width	Length	Square Inches	Square Feet

Bill of Materials

Project Name: _____

Quantity	Units	Material Description	Price/Unit	Extended Price

			Subtotal	
			Tax	
			Total	

Plan of Procedure

Minutes, Hours, Days, Weeks, or Months

Steps	1	2	3	4	5	6	7	8	9	10	11	12	13	14	15	16	17	18	19	20	21	22	23	24	25
1																									
2																									
3																									
4																									
5																									
6																									
7																									
8																									
9																									
10																									
11																									
12																									
13																									
14																									
15																									
16																									
17																									
18																									
19																									
20																									
21																									
22																									
23																									
24																									
25																									
26																									
27																									
28																									
29																									
30																									
31																									
32																									

Material Cut Sheet

Part Name	Material	Number Pieces	Rough Measurements			Cut Process Completed	Finish Measurements			Cut Process Completed
			Thickness	Width	Length		Thickness	Width	Length	

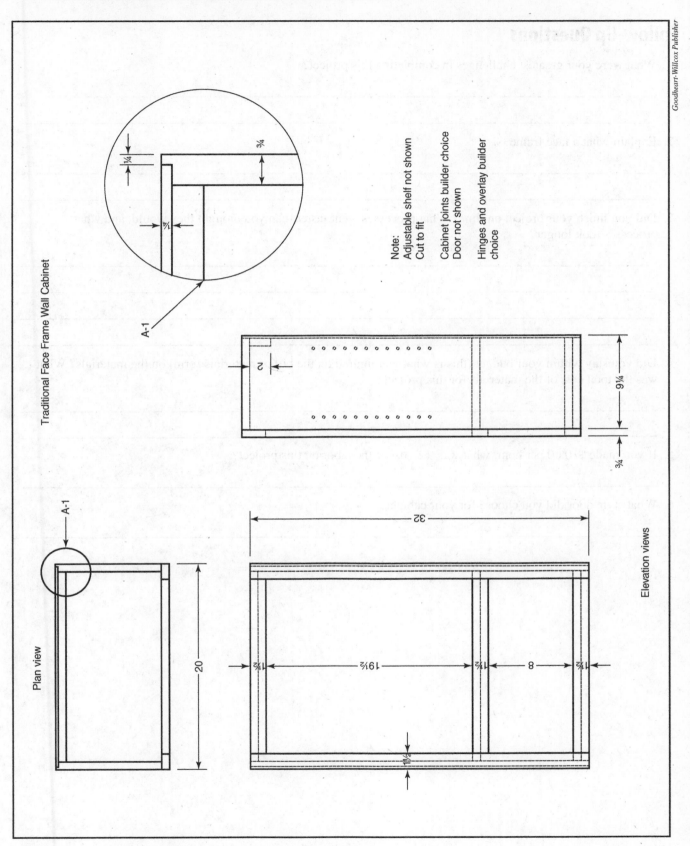

Traditional Face Frame Wall Cabinet

¾

¼

½

A-1

Note:
Adjustable shelf not shown
Cut to fit

Cabinet joints builder choice
Door not shown

Hinges and overlay builder
choice

2

9¼

¾

A-1

Plan view

20

32

1½

19½

1½

8

1½

1½

Elevation views

Follow-Up Questions

1. What were your greatest challenges in completing this project?

   _____

   _____

2. Explain what a face frame is.

   _____

   _____

3. Did you finish your project on time? What processes went faster than you thought they would, and what processes took longer?

   _____

   _____

   _____

   _____

4. Did you stay within your budget (this is what you figured on the bill of materials form) on the materials? What was the total cost of the materials for this project?

   _____

   _____

5. If you made $10.00 per hour, what was the cost of the labor on this project?

   _____

6. What style door did you choose for your cabinet?

   _____

Rubric for Wall Cabinets—Face Frame and Carcase Construction

Student Guidelines

Materials and resources

Possible points _____   Score _____

Did you complete a bill of materials, complete a list of resources, and figure total cost of materials?

_____

Cut list

Possible points _____   Score _____

Is the 4′ × 8′ form and cut list filled out correctly and completely?

_____

Plan of procedure

Possible points _____   Score _____

Did you keep a detailed list of steps to build the project by properly completing the Plan of Procedure sheet? Did you keep accurate time records of estimated time and actual time for each step?

_____

Craftsmanship

Possible points _____   Score _____

Did you follow plans, sand out mill marks, and ensure corners are broken properly? Are edges flush at joints? Are measurements accurate, and the cabinet square?

_____

Follow-up questions

Possible points _____   Score _____

Were all questions answered in a thoughtful manner?

_____

Total possible points _____   Score _____

Section Project 5-5

Frameless Bedside Stand Cabinets

This project is designed to give you practice in the basic frameless method of constructing cabinets. It will feature a solid full overlay mount door and recessed back. There are options for door design and drawer construction. Deviation from the plans should be backed up by your shop drawings. Another optional challenge is to modify the cabinet for the 32mm System.

Resource Chapters

Chapter 9—Production Decisions
Chapter 22—Sawing with Hand and Portable Power Tools
Chapter 23—Sawing with Stationary Power Machines
Chapter 34—Overlaying and Inlaying Veneer
Chapter 37—Joinery
Chapter 40—Case Construction
Chapter 41—Frame and Panel Construction
Chapter 43—Doors

Objectives

After completing this project, you should be able to:

- Demonstrate skill in carcase construction.
- Master edgebanding techniques.
- Install European concealed hinges.
- Cut cabinet parts from 4′ × 8′ sheets.
- Construct drawers.
- Use all equipment and tools of the shop to the task of building cabinets.

Materials and Resources

- Cabinet grade 3/4″ 4′ × 8′ MDF, plywood, or melamine
- Edgeband material
- Drawer stock
- Drawer guides
- European concealed hinges
- Finish supplies and sandpaper
- Shelf pins

> **SAFETY NOTE**
>
> Before proceeding with this activity, you should have passed a general shop safety rule test, all specific safety tests on the tools used in this activity, and be certified by the instructor to use the tools and equipment.

Activity Procedure

1. Study the plans.

2. Make a cut list.

3. Make a bill of materials.

4. Make a list of resources needed other than materials, such as tools and equipment.

5. Before making the plan of procedure, a decision must be made regarding the type of joints, edgeband, and drawer glides to be used. Your shop capabilities and instructor's preference will influence these decisions.

6. Make a plan of procedure. Refer to Chapter 9, Sections 9.3 and 9.4.

7. Execute your plan of procedure.

Final Project Requirements

- Bill of materials
- 4′ × 8′ layout plan and cut list
- Material cut sheet
- Plan of procedure
- Shop sketches of door design
- Finished frameless bedside stand
- Follow-up questions

4′ × 8′ Layout Plan and Cut List

Name or Job _____ Period _____ Date _____

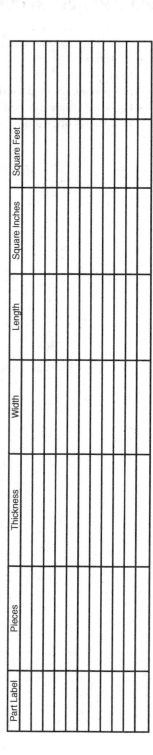

Part Label	Pieces	Thickness	Width	Length	Square Inches	Square Feet

Bill of Materials

Project Name: _____

Quantity	Units	Material Description	Price/Unit	Extended Price

	Subtotal	
	Tax	
	Total	

Material Cut Sheet

Part Name	Material	Number Pieces	Rough Measurements			Cut Process Completed	Finish Measurements			Cut Process Completed
			Thickness	Width	Length		Thickness	Width	Length	

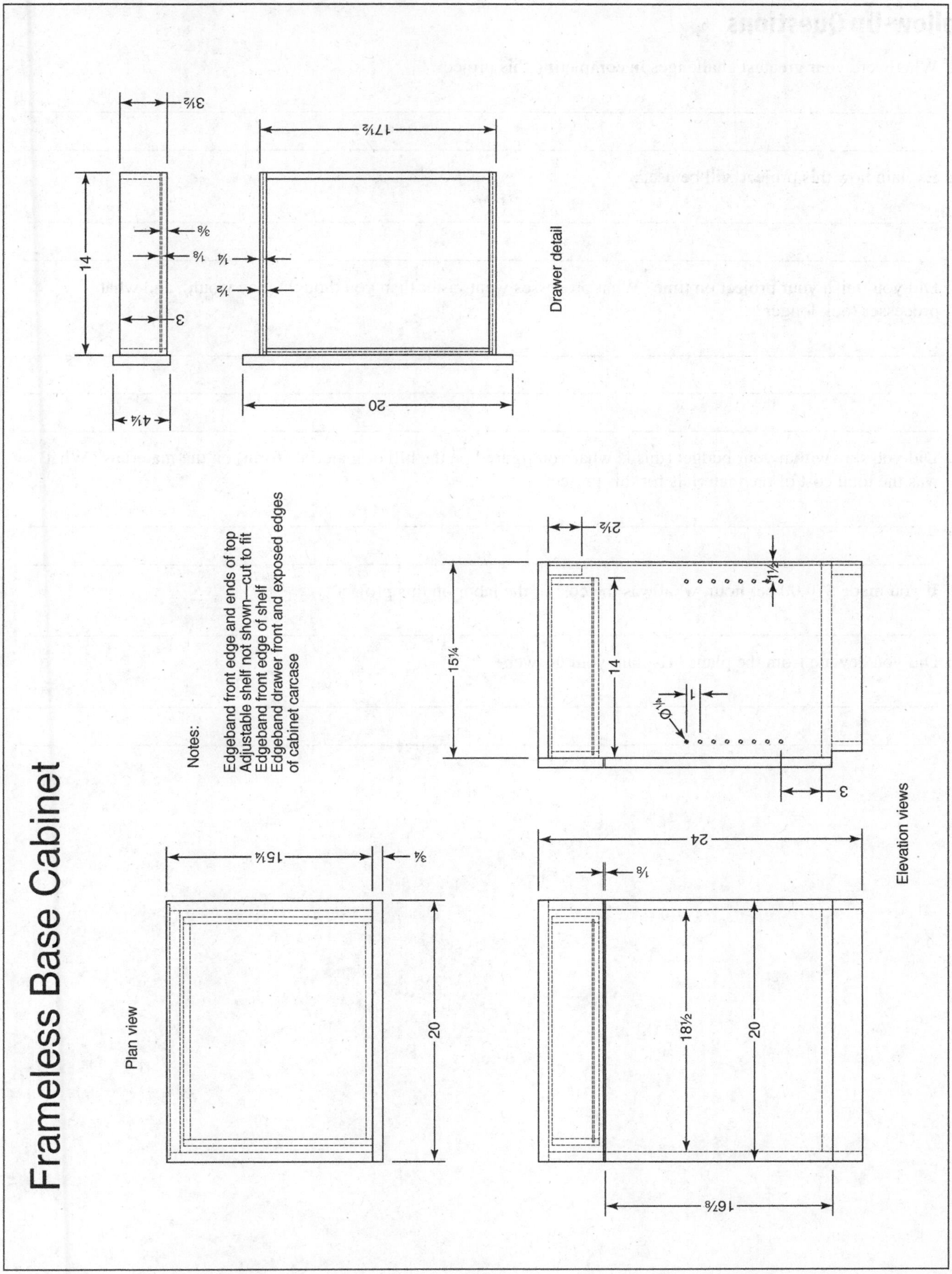

Goodheart-Willcox Publisher

Frameless Base Cabinet

Notes:
Edgeband front edge and ends of top
Adjustable shelf not shown—cut to fit
Edgeband front edge of shelf
Edgeband drawer front and exposed edges
of cabinet carcase

Drawer detail

3½

17½

14

⅜
⅛
¼
½
3

4¼

20

Plan view

15¼

¾

20

Elevation views

2½

1½

15¼

14

1
¼ Ø

3

24

⅛

18½

20

16⅞

Follow-Up Questions

1. What were your greatest challenges in completing this project?

   _____

   _____

2. Explain how this project will be used.

   _____

   _____

3. Did you finish your project on time? What processes went faster than you thought they would, and what processes took longer?

   _____

   _____

   _____

4. Did you stay within your budget (this is what you figured on the bill of materials form) on the materials? What was the total cost of the materials for this project?

   _____

   _____

5. If you made $10.00 per hour, what was the cost of the labor on this project?

   _____

6. Did you deviate from the plans? Explain your answer.

   _____

   _____

Rubric for Frameless Bedside Stand Cabinets

Student Guidelines

Plan of procedure

Possible points _____ Score _____

Did you properly use the Process Plan sheet to make a list of the steps and estimate the time required to complete each task?

_____

Cut list

Possible points _____ Score _____

Is the 4′ × 8′ form and cut list filled out correctly and completely?

_____

Materials and resources

Possible points _____ Score _____

Did you complete a bill of materials, complete a list of resources, and figure total cost of materials?

_____

Craftsmanship

Possible points _____ Score _____

Did you follow plans, sand out mill marks, and ensure corners are broken properly? Are edges flush at joints? Are measurements accurate, and the cabinet square?

_____

Follow-up questions

Possible points _____ Score _____

Were all questions answered in a thoughtful manner?

_____

Total possible points _____ Score _____

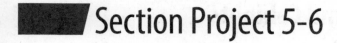 Section Project 5-6

Frameless Wall Cabinets

This project is designed to give you practice in the basic frameless method of constructing cabinets. It will feature a solid flush mount door, recessed back. Butt joints can be biscuit joined, doweled, or be finish nailed.

Resource Chapters

Chapter 9—Production Decisions
Chapter 22—Sawing with Hand and Portable Power Tools
Chapter 23—Sawing with Stationary Power Machines
Chapter 34—Overlaying and Inlaying Veneer
Chapter 37—Joinery
Chapter 40—Case Construction
Chapter 41—Frame and Panel Construction
Chapter 43—Doors

Objectives

After completing this project, you should be able to:

- Explain the procedure to construct a simple frameless European-style cabinet.
- Compare the work and expertise needed to build a frameless cabinet to building a face frame cabinet.

Materials and Resources

- Cabinet grade 3/4″ and 1/4″ panel materials
- Suitable edgebanding
- European concealed hinges with full overlay plates
- Finish supplies and sandpaper
- 5 mm shelf pins

SAFETY NOTE

Before proceeding with this activity, you should have passed a general shop safety rule test, all specific safety tests on the tools used in this activity, and be certified by the instructor to use the tools and equipment.

Activity Procedure

1. Study the plans.

2. Make a cut list.

3. Make a bill of materials.

4. Make a list of resources needed other than materials, such as tools and equipment.

5. Before making the plan of procedure, a decision must be made regarding the type of joints and edgeband to use. Shop capabilities and your instructor's preference will influence these decisions.

6. This project could be used to practice applying plastic laminate or veneer. Discuss these possibilities with your instructor before filling out the plan of procedure.

7. Make a plan of procedure. Refer to Chapter 9, Sections 9.3 and 9.4.

8. Execute your plan of procedure.

Final Project Requirements

- Bill of materials
- Plan of procedure
- $4' \times 8'$ layout plan and cut list
- Material cut sheet
- Finished frameless wall cabinet
- Follow-up questions

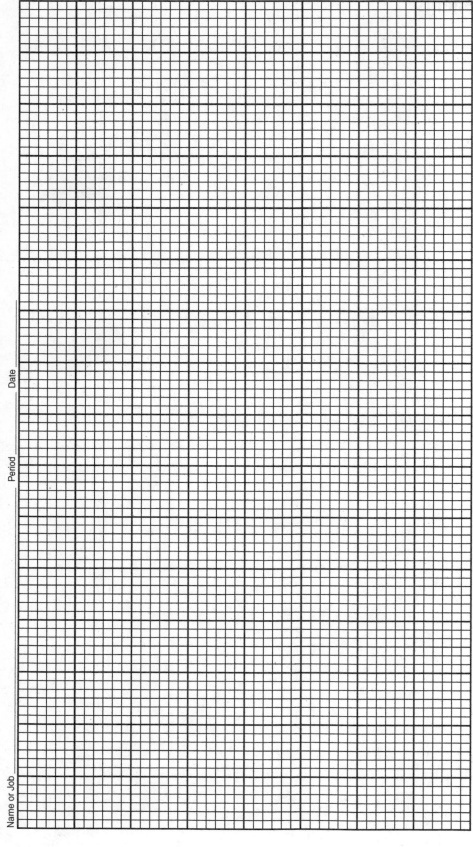

4' × 8' Layout Plan and Cut List

Name or Job _____ Period _____ Date _____

Part Label	Pieces	Thickness	Width	Length	Square Inches	Square Feet

Name _____

Bill of Materials

Project Name: _____

Quantity	Units	Material Description	Price/Unit	Extended Price

Subtotal	
Tax	
Total	

Material Cut Sheet

Part Name	Material	Number Pieces	Rough Measurements			Cut Process Completed	Finish Measurements			Cut Process Completed
			Thickness	Width	Length		Thickness	Width	Length	

Frameless Wall Cabinet

3/4

1/4

1/4

A-1

Note:
Adjustable shelf not shown
Edge trim all exposed edges
Cabinet joints builder choice
Double or single door builder
choice

10

2

7¾

1

4¾

1¼

8⅞

10

A-1

Plan view

20

32

9¹⁵/₁₆

Elevation views

Follow-Up Questions

1. What were your greatest challenges in completing this project?

   _____

   _____

   _____

2. Explain how this project will be used.

   _____

   _____

   _____

3. Did you finish your project on time? What processes went faster than you thought they would, and what processes took longer?

   _____

   _____

   _____

4. If you made $10.00 per hour, what was the cost of this project?

   _____

5. Did you stay within your budget (this is what you figured on the bill of materials form) on the materials? What was the total cost of the materials for this project?

   _____

   _____

Name _____

Rubric for Frameless Wall Cabinets

Student Guidelines

Materials and resources

Possible points _____ Score _____

Did you complete a bill of materials, the 4′ × 8′ layout plan and cut list, and figure total cost of materials?

_____

Plan of procedure

Possible points _____ Score _____

Did you keep a detailed list of steps to build the project? Did you keep accurate time records of estimated time and actual time for each step?

_____

Craftsmanship

Possible points _____ Score _____

Did you follow plans, sand out mill marks, and ensure corners are broken properly? Are edges flush at joints? Are measurements accurate, and the cabinet square?

_____

Follow-up questions

Possible points _____ Score _____

Were all questions answered in a thoughtful manner?

_____

Total possible points _____ Score _____

Notes

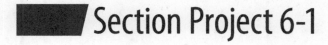 Section Project 6-1

Fixing a Dent

While cabinetmakers strive not to make unnecessary scratches and dents in their projects, sometimes mistakes happen. Therefore, the cabinetmaker should be skilled at fixing these mistakes.

Resource Chapter

Chapter 51—Preparing Surfaces for Finish

Objective

After completing this project, you should be able to:

- Use a household iron to remove a dent.

Materials and Resources

- Household iron
- 4″ × 4″ wood scraps
- 120 grit sandpaper

> **SAFETY NOTE**
>
> Before proceeding with this activity, you should have passed a general shop safety rule test, all specific safety tests on the tools used in this activity, and be certified by the instructor to use the tools and equipment.

Activity Procedure

1. Find a small piece of soft maple, pine, or poplar and cut to 4″ × 4″.
2. Use a bell-faced hammer to dent the wood.
3. Place a 16d nail laid flat on the wood and hit the nail to put a small dent in the wood.
4. In both cases, try to not cut the fibers, but only create a dent.
5. Soak the dented areas with warm water.
6. Being careful not to touch the hot surface or lay it next to flammable materials, place the iron on the dent. The water should steam the dent out.
7. Repeat the process if your first attempt does not remove the dent.
8. Lightly sand the piece to determine if the dent completely disappears.

Final Project Requirements

- Wood scrap with dent removed
- Follow-Up Questions

Follow-Up Questions

1. What did you learn from this activity?

_____

_____

_____

_____

2. Did you burn the wood with the iron? If so, how do you think you can avoid this from happening the next time?

_____

_____

_____

_____

3. At what point in the cabinetmaking process do you think dent removal should take place?

_____

_____

_____

_____

Name _____

Rubric for Fixing a Dent

Student Guidelines

Dent removal
Did you successfully remove the dent?

Possible points _____   Score _____

_____

Follow-up questions
Were all questions answered in a thoughtful manner?

Possible points _____   Score _____

_____

Total possible points _____   Score _____

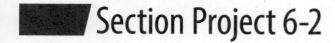 Section Project 6-2

Finish Sample Set

The main function of finish is to protect and beautify a wood or metal product. The purpose of this activity is to become familiar with the many finishes used in your shop. In this activity, you will make a sample set of those finishes.

Resource Chapters

Chapter 50—Finishing Decisions
Chapter 51—Preparing Surfaces for Finish
Chapter 52—Finishing Tools and Equipment
Chapter 53—Stains, Fillers, Sealers, and Decorative Finishes
Chapter 54—Topcoatings

Objective

After completing this project, you should be able to:

- Demonstrate an understanding of the finishing methods used in your shop.

Materials and Resources

- 1/4″ solid wood or plywood
- Finishing supplies used to finish cabinets in your shop
- Plan of procedure sheet

> **SAFETY NOTE**
> - Make sure when applying solvent-based finishes to work in a well-ventilated area and to use proper respirators.
> - Before proceeding with this activity, you should have passed a general shop safety rule test, all specific safety tests on the tools used in this activity, and be certified by the instructor to use the tools and equipment.

Activity Procedure

1. Cut three to six 1/4″ × 3″ × 6″ wood samples (one sample of each wood for each finishing recipe).

2. Cut a 1/16″ deep saw kerf across the grain of the pieces, 2″ from the end.

3. Finish sand each piece.

4. Write your finishing procedures in the plan of procedure sheet.

5. Place a piece of masking tape on the edge of the saw kerf to mask off the 2″ side.

6. Perform your finishing procedures on the 4″ portion of the piece, leaving the 2″ portion unfinished.

7. Repeat for each sample.

Final Project Requirements

- Finished samples
- Plan of procedure sheet with finishing procedure (time section can be left blank)
- Follow-Up Questions

Plan of Procedure

Minutes, Hours, Days, Weeks, or Months

Steps	1	2	3	4	5	6	7	8	9	10	11	12	13	14	15	16	17	18	19	20	21	22	23	24	25
1																									
2																									
3																									
4																									
5																									
6																									
7																									
8																									
9																									
10																									
11																									
12																									
13																									
14																									
15																									
16																									
17																									
18																									
19																									
20																									
21																									
22																									
23																									
24																									
25																									
26																									
27																									
28																									
29																									
30																									
31																									
32																									

Follow-Up Questions

1. What were your greatest challenges in completing this project?

_____

_____

_____

_____

2. Explain how this project will be used.

_____

_____

_____

_____

3. What thinners were used in your finishing procedures?

_____

_____

_____

_____

4. What methods did you use to apply your finishes? Explain.

_____

_____

_____

_____

Name _____

Rubric for Finish Sample Set

Student Guidelines

Required number of samples
Possible points _____ Score _____

Did you complete the number of samples set by the instructor?

_____

Followed directions
Possible points _____ Score _____

Did you follow correct procedures for removing and storing supplies and for cleanup? Were rags disposed of properly and brushes and equipment cleaned?

_____

Plan of procedure
Possible points _____ Score _____

Did you fill out the sheet in detail?

_____

Craftsmanship and neatness
Possible points _____ Score _____

Were the finishes executed correctly?

_____

Follow-up questions
Possible points _____ Score _____

Were all questions answered in a thoughtful manner?

_____

Total possible points _____ Score _____